STP 1411

AF859680

Applications of Automation Technology in Fatigue and Fracture Testing and Analysis: Fourth Volume

A. A. Braun, P. C. McKeighan, A. M. Nicolson, and R. D. Lohr, editors

ASTM Stock Number: STP 1411

ASTM
100 Barr Harbor Drive
PO Box C700
West Conshohocken, PA 19428-2959

Printed in the U. S. A.

ISBN: 0-8031-2890-8
ISSN: 1537-7407

Peer Review Policy

Each paper published in this volume was evaluated by two peer reviewers and at least one editor. The authors addressed all of the reviewers' comments to the satisfaction of both the technical editor(s) and the ASTM Committee on Publications.

To make technical information available as quickly as possible, the peer-reviewed papers in this publication were prepared "camera-ready" as submitted by the authors.

The quality of the papers in this publication reflects not only the obvious efforts of the authors and the technical editor(s), but also the work of the peer reviewers. In keeping with long-standing publication practices, ASTM maintains the anonymity of the peer reviewers. The ASTM Committee on Publications acknowledges with appreciation their dedication and contribution of time and effort on behalf of ASTM.

Printed in Chelsea, MI
January 2002

Foreword

This publication, *Applications of Automation Technology in Fatigue and Fracture Testing and Analysis: Fourth Volume,* contains papers presented at the symposium of the same name held in Orlando, FL, on 15 November 2000. The symposium was sponsored by ASTM Committee E8 on Fatigue and Fracture. The symposium co-chairmen were Arthur A. Braun, MTS Systems Corporation, Peter C. McKeighan, Southwest Research Institute, Murray Nicolson, Instron Corporation, and Raymond Lohr, Instron Ltd.

Contents

Overview

The greatest technological gain that has occurred in the mechanical testing laboratory in the past twenty years arguably has been the benefits as a result of the persistent and rapid growth of computer technology. Although sensor technology has also evolved considerably over this time, the new features that have resulted with higher performance, low cost hardware, and software systems are providing exciting new capability in the general areas of test control, data acquisition, data analysis and interpretation, modeling, and integration of testing and design.

This symposium is the fourth in a series of symposia concerned with advancing the state of the art in automated fatigue and fracture testing. This series of meetings was initiated in 1975 with STP 613, entitled "Use of Computers in the Fatigue Laboratory" and held in New Orleans, Louisiana in November, 1975. Although it is hard to believe, the personal computer as we know it was still five years away when the first symposia was held in 1975. Over the past two and a half decades, the role of the computer in the test laboratory has dramatically altered the range of test control and analysis capabilities available.

For example, purchasing a servohydraulic test system today typically includes a digital control system to provide an interface between the user and the control of the frame. Although analog controllers can be purchased, the clear trend for the future is digital command and control. Twenty-five years ago, it was the exception rather than the rule to see a computer attached to a servohydraulic test machine. This is contrasted by today's mechanical test laboratory, where it is not uncommon to see *multiple* personal computers connected to the same test frame, where one might be controlling the test and the second involved in highly specialized data acquisition.

The rapid changes in computer technology have created some problems with regard to the stability of tools in the laboratory. As an example of this, consider one of the latest trends of personal computers where the DOS operating system is no longer accessible. The tools developed during the 1980s and early 1990s were written based on this platform. The absence of DOS means that some applications that work perfectly well can no longer be used with modern hardware. This software-retirement-through-hardware-obsolescence is an issue that needs to be further examined and worked on to minimize extra expense. This example is not the only occurrence of this; component level (e.g., cards and chips) hardware nonavailability has also impacted "the big boys," as some of the servohydraulic system manufacturers have had to accelerate software development to accommodate obsolete hardware.

Given this computer development and its growing role in the test laboratory, the question that can be asked is what do we really do differently today, as opposed to the precomputer days. Without question, tests have become more automatic and, by virtue of this, more efficient to run. As an example of this, in the precomputer days fatigue crack growth tests were laborious efforts with a technician spending considerable time staring down a microscope. Today, a test can virtually be started at the end of the day shift and the results be available the next morning. Whilst this has become more efficient, coping with the vast quantities of data that can be generated can be overwhelming. Automated tools for performing analysis are continually evolving to provide the test engineer with the critically required quantity from his transducer data.

The test engineer is faced with a challenge to attempt to keep technical knowledge current with the continual developmental onslaught that occurs with modern silicon devices. This symposium, and the fourteen papers presented, provides some bases to understand the range of applications that computers have in the modern test lab. Classifying the content of the papers included is difficult, since the range is quite broad. Nevertheless, a number of papers examine the challenges faced in full-scale testing, either from a control or end-level editing viewpoint. Several papers also examine how fatigue or fracture data are applied in the design process to yield safer structures with longer service lives. As described, a variety of computer-based lifting tools are now available to users to apply to the design process. Finally, a number of papers examined specific system implementations, especially as related to more challenging applications such as high frequency or thermomechanical fatigue testing. The applications undertaken in the latest reported systems with the newest automated testing software include some of the greatest testing challenges currently faced in the mechanical testing laboratory. This is certainly a new development as the computer and software each have increased capability, speed, and flexibility.

In summary, this symposium and the proceedings herein are intended to provide an update on the applications of automation in the fatigue and fracture testing laboratory. It is the intention of the Automation Task Group in ASTM E08 to revisit this area every three or four years to report and track how testing evolves. This is a developmental area that will continue to flourish as technologists apply the newer, faster, and bigger hardware, and software engineers create the newest generation of data manipulation tools.

Finally, the editors would like to express their sincere appreciation to all the authors and co-authors responsible for the papers included in this STP and the presentations made during the symposium. Furthermore, we would like to recognize the efforts of the reviewers whose high degree of professionalism and timely response ensure the quality of this publication. Finally, the editors would also like to express their sincere gratitude to the ASTM planning and editorial staff for their assistance with the symposium, as well as their critical input to this special technical publication.

Peter C. McKeighan
Southwest Research Institute
San Antonio, Texas
Symposium co-chairman and co-editor

Systems Implementations

Claude Bathias,[1] J. M. De Monicault,[2] and G. Baudry[3]

Automated Piezoelectric Fatigue Machine for Severe Environments

Reference: Bathis, C., De Monicault, J. M., and Baudry, G., "**Automated Piezoelectric Fatigue Machine for Severe Environments**," *Applications of Automation Technology in Fatigue and Fracture Testing and Analysis: Fourth Volume, ASTM STP 1411*, A. A. Braun, P. C. McKeighan, A. M. Nicolson, and R. D. Lohr, Eds., American Society for Testing and Materials, West Conshohocken, PA, 2002.

Abstract: During the 1990 s several methods have been developed around the world in order to test specimens at very high fatigue life (for example SWRI, Air Force Laboratory in the US, the University of Vienna in Europe, and NRIM in Japan). In our laboratory an automatic ultrasonic fatigue testing system was designed and built 10 years ago to determine the fatigue crack growth threshold of metallic alloys. Those first results were published in ASTM STP 1231 in 1994. Since this date, many applications of this device were made facing different technological challenges.

At this time our machine is working at 20kHz, with R ratio between −1 and 0.8, at room temperature, high temperature, cryogenic temperature, atmospheric pressure, and high pressure up to 300 bar. The system was designed for special applications such as testing in a hydrogen gas, hydrogen liquid or water or salt water, and to determine SN curves up to 10^{10} cycles.

Keywords: piezoelectric machine, gigacycle fatigue, environmental effects, cryogenic temperature, fretting fatigue

It is interesting to point out that many structural components are working beyond 10^7 cycles facing severe environments such as temperature, wear or corrosion, that is to say, in the gigacycle fatigue regime.

From an historical point of view, it is said that the first ultrasonic fatigue machine was constructed in 1950 by Mason *[1]* and it was the beginning of the discovery of gigacycle fatigue. With the development of computer techniques, C. Bathias and co-workers [2-4] have recently built a fully computer controlled piezoelectric fatigue machine working at 20kHz ± 0.5 kHz. The vibration of the specimen is induced with a piezo-ceramic transducer, which generates an acoustical wave to the specimen through a power concentrator (horn) in order to obtain more important displacement and an amplification of the stress. The resonant length of the specimen and concentrator is calculated using FEM. In our machine, there is a linear relation between the electric potential and the dynamic displacement amplitude of the

[1] Professor, CNAM-ITMAA, 2 rue Conte, 75003 Paris, France
[2] Engineer, SNECMA, Foret de Vernon, 27207 Vernon, France
[3] Engineer, ASCOMETAL, 57301 Hagondange, France

ceramic in order to keep the stress constant, during the test, via computer control.

The test is automatically stopped when the frequency falls below 19.5 kHz. The basic machine and specimens are described in others papers [2]. It must be noticed that this machine is not operative below a fatigue life of 10^6 cycles because elasto-plasticity becomes higher and higher.

At this time, our piezoelectric fatigue systems are working at 20kHz, with R ratio between –1 and 0.8 at room temperature, high temperature, cryogenic temperature, atmospheric pressure, high pressure and fretting-fatigue. For special applications this piezoelectric fatigue machine is able to test specimens in severe environments such as hydrogen gas, hydrogen liquid, to determine SN curves up to 10^9 cycles

In this paper, variants of this piezoelectric fatigue system are presented, including computer control, computerized data acquisition and computerized generation of test results.

Cryogenic Temperature

The device consists of three parts: a cryostat, a mechanical vibrator and a controlled power generator. Figure 1 shows the principal aspect of this machine; it is simpler than a conventional hydraulic machine. In this apparatus, the converter changes an electronic signal into a mechanical vibration; the horn plays the role of amplitude amplifier. A cryostat contains cryogenic liquid to maintain a constant testing temperature (Fig. 2).

A generator with a converter consisting of six piezo-ceramics was chosen to provide vibration energy. The converter, horn and specimen compose a mechanical vibration system where there are four stress nodes (null stress) and three displacement nodes (null displacement) for an intrinsic frequency (20 kHz). Here, the stress and displacement are defined as longitudinal stress and displacement because the structure is relatively long. In Fig. 1, points B, C (connected points), point A and converter top are stress nodes. The specimen center is a displacement node, but the stress is maximum.

The horn has to be calculated to vibrate at a frequency of 20 kHz. Depending on the specimen loading, the horn is designed to get an amplification of the displacement amplitude between B and C usually from 3 and 9. It means that the geometry between B and C can be modified (Fig. 1). The finite element method may be used when the geometrical shape is complex.

The key points of the machine are given below:

1. The mechanical system composed of a converter, a horn and a linear specimen, since all stress and displacement fields are linear.
2. Only displacement is needed to determine the stress field.
3. To avoid the use of a load sensor, the stress in the mid-section of the specimen is computed from the displacement of the piezo-ceramics system.

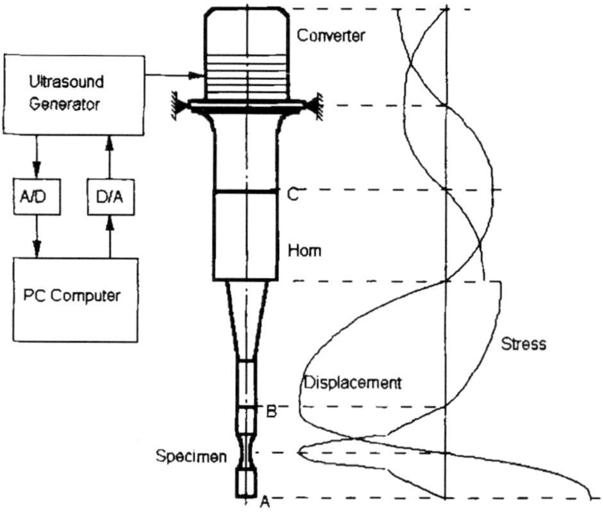

Figure 1 – *Vibratory stress and displacement field, and computer control system*

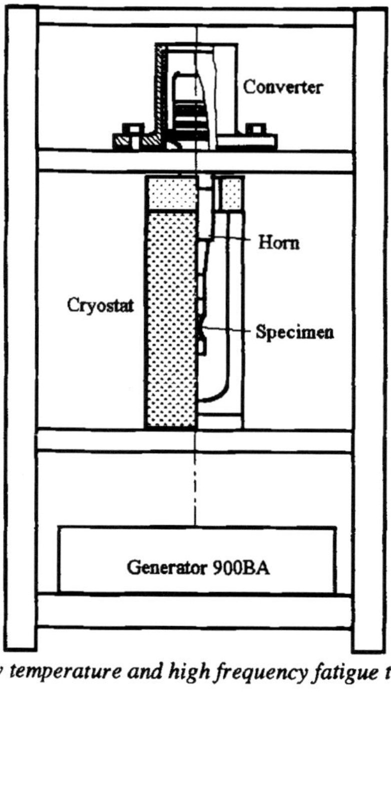

Figure 2 – *Low temperature and high frequency fatigue testing machine*

The piezo-ceramics expand or contract when an electrical field is applied. The voltage is proportional to expansion or contraction, i.e. the voltage is proportional to the displacement in the mechanical system. It is strictly proportional to expansion or contraction of the converter and to the displacement of the point C. That is, electrical current depends on the damping of the horn and specimen installed on the converter. In the generator, an interface called J2 has been set up, in which there is a plug giving 0-10 volts DC corresponding to 0-100% of vibration amplitude of the converter. This output is calibrated with the displacement of the horn end (point B), to determine the stress in the specimen using a computer that acquires this voltage. The stress can be calculated by the following equation (1):

$$\sigma = Ek_s k_h U_{c100\%} \frac{V}{10}, \qquad (1)$$

where E is Young's modulus k_s is a factor of the specimen dependent on geometrical form, k_h is the ratio of amplitude amplification, $U_{c100\%}$ is maximum amplitude at point C which is constant and V is DC tension acquired by the computer. According to this formula, the test stress for a certain specimen can be modified not only by changing output power but also by replacing the horn.

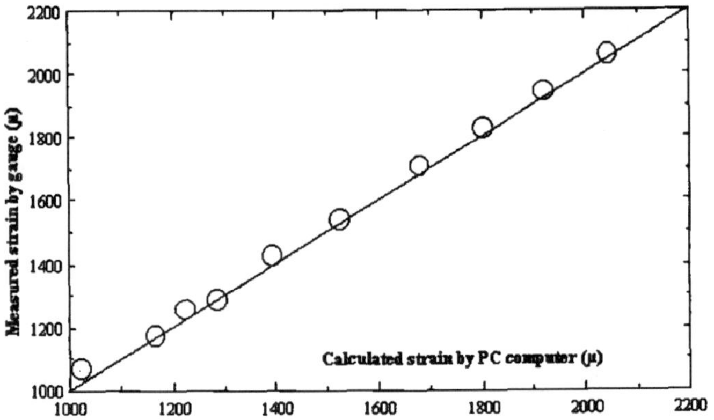

Figure 3 – *Comparison of results of measured strain and calculated strain at 77 K*

For calibration, a simple cylindrical specimen was used, whose center was instrumented by a strain gauge. Measured strain (e) by this gauge and displacement of horn end at B U_B is calculated by the following relation (2):

$$e = 2\pi f\, U_B \sqrt{\frac{P}{E}} \qquad (2)$$

where f is frequency, and P is density. When the DC output is calibrated according to this measurement, a comparison between measured strain in liquid nitrogen and calculated strain by computer control for different power can be presented in Fig. 3. It is seen that the linearity is good, and that error between measured and calculated values is small.

Other calibration tests have been performed by using an optical sensor to measure displacement of the specimen at room temperature. It is possible to apply a correction from room temperature to lower temperature since the amplification ratio is known for different temperatures. The results are also satisfactory.

In the interface J2, there is another plug to which a DC voltage of 0-10 volts can be given to control vibration amplitude. In general, direct control at 20kHz is very difficult. Thus, it is more reliable to use direct current signal proportional to amplitude of alternating current signal [4]. A normal A/D and D/A converter card connecting the connector J2 and a PC can enable a computer to control tests at 20 kHz. Such a control program has been written in Turbo C + +. It calculates the vibration stress in the specimen for various materials. The test starts by giving a target test stress, and the real stress rises within 85 milli-seconds to the expected level without overloading. Then, the stress is held constant and control accuracy is
± 10 Mpa. When a crack appears, the testing system stops automatically because of decreasing frequency and it thus measures the fatigue life for a frequency drop of 2.5% the crack length is of the order of one millimeter. Owing to this software, fatigue tests between 10^5 to 10^{10} cycles can be performed.

In Fig. 4 it can be seen that fatigue lives of titanium alloys are scattered and that the results of vibratory fatigue and conventional fatigue are coherent. Nevertheless, a small difference is observed between two SN curves at 20 kelvin, since one is obtained in liquid hydrogen and the other one in gaz helium. It could be related to the temperature control inside the cryostat. Generally, titanium alloy fatigue behavior is better at cryogenic temperature than at room temperature. In addition, fractographic examination did not show special phenomena in high frequency fractured specimens.

Other tests have been carried out for titanium alloy Ti6246 to determine the fatigue strength at 10^9 cycles at 77 K with this machine. The results are shown in (Fig.5.) In these experiments, three microstructures were produced from different thermal processing procedures. We can see that *S-N* curves range between 10^7 and 10^9 cycles. It appears to be a large effect of the thermal processing. The lowest fatigue strength of the C material is explained by large primary alpha platelets due to slow solution treatment. The best fatigue strength at 77 K is obtained with a fine microstructure. In all cases, it is shown that the SN curve does not present any asymptot between 10^6 and 10^9 cycles at cryogenic temperature.

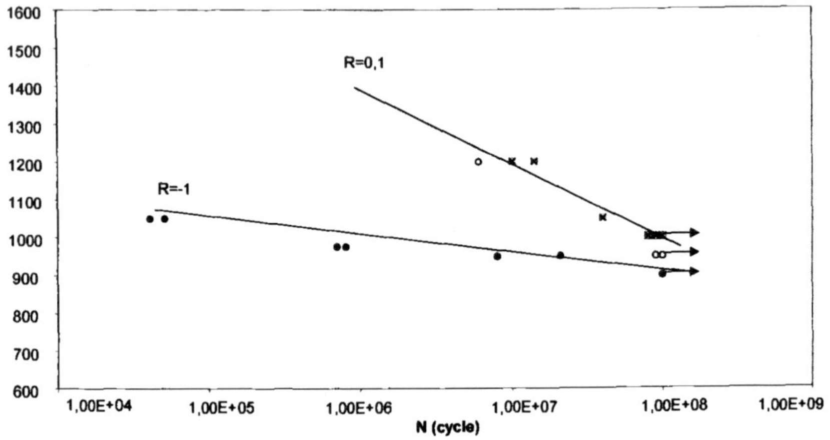

Figure 4 – *Titanium alloy in hydrogen liquid (R=-1) and helium at 20 Kelvin*

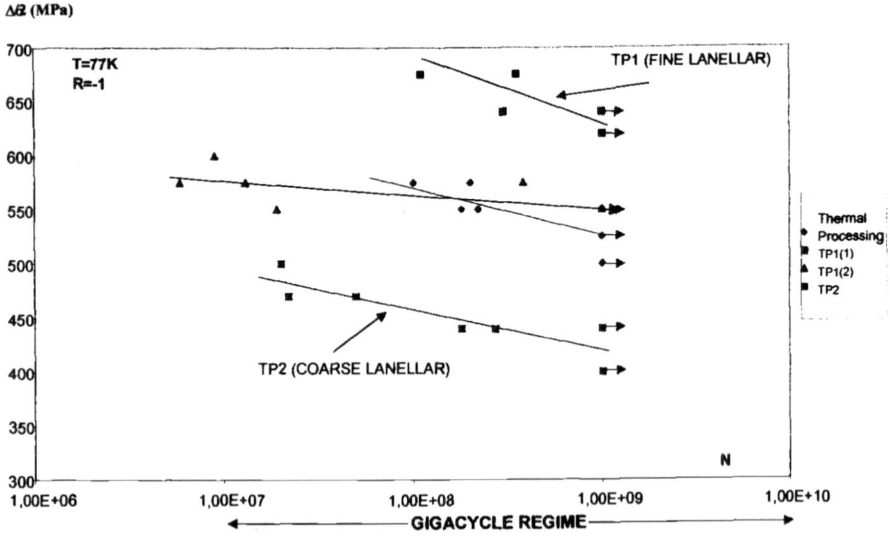

Figure 5 – *Gigacycle fatigue of Ti-6246 at 77Kelvin*

High Temperature Testing

A schematic view of the piezoelectric fatigue machine is shown in (Fig.1). The specimen is heated with an inductive coil in order to get a constant temperature from 400 to 800°C along a 20 mm gage length.

Figure 6 presents some results of high frequency tests with an R = -1 for a powder metal N18 alloy. One can see that the threshold is smaller at high temperature than at ambient temperature. Normally we would expect decreasing threshold with an increase in temperature. But in Figure 6 the threshold is smaller at 400°C than at 650°C and 750°C. The curves at 400°C, 650°C and 750°C cut the vicinities to 10^5 mm/cycle. The observed gaps are explained by the phenomenon of oxidization at the bottom of the crack. On the crack surface of the samples used in our tests, oxidization at 650°C and at 750°C was observed. At high temperature, crack propagation rate normally increases with the temperature but the oxidization could slow propagation down in the threshold range to a small load when the temperature is rather elevated. The same phenomenon is observed at low frequency. Thus, it seems that the effect of corrosion is similar at 20kHz and at low frequency.

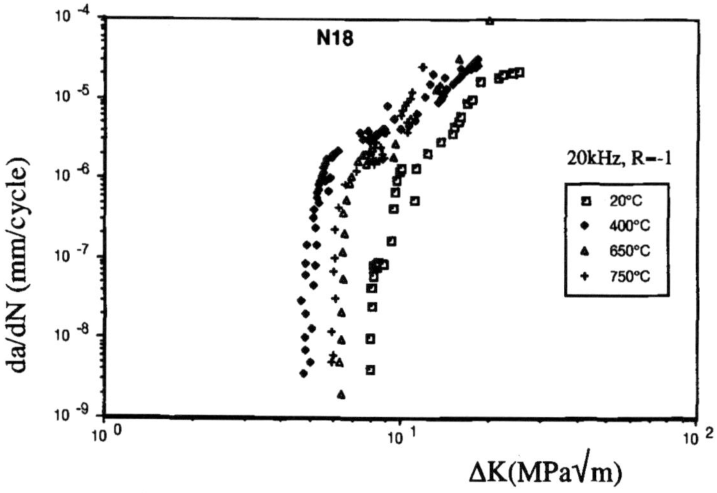

Figure 6 – *Fatigue threshold for N18 at 450°C*

High Pressure Piezo-Electric Fatigue Machine

It is well known that it is difficult to carry out a fatigue tests under high pressure with a conventional machine. The problem stems from the displacement of an actuator through the wall of an autoclave. Using a piezo-electric fatigue system this problem disappears because it is easy to get zero displacement at the location where the sonotrode is crossing the wall of the autoclave.

Thus a high pressure piezoelectric fatigue machine for testing in pressures up to 300 bar has been built in our laboratory. The design is shown in Figure 7.

With this device, it has been shown that hydrogen under a pressure of 100 bar has an effect on the SN curve of IN 718 at room temperature. In Figure 8 two SN curves in hydrogen and in helium are compared in order to show the hydrogen effect between 10^6 and 10^9 cycles.

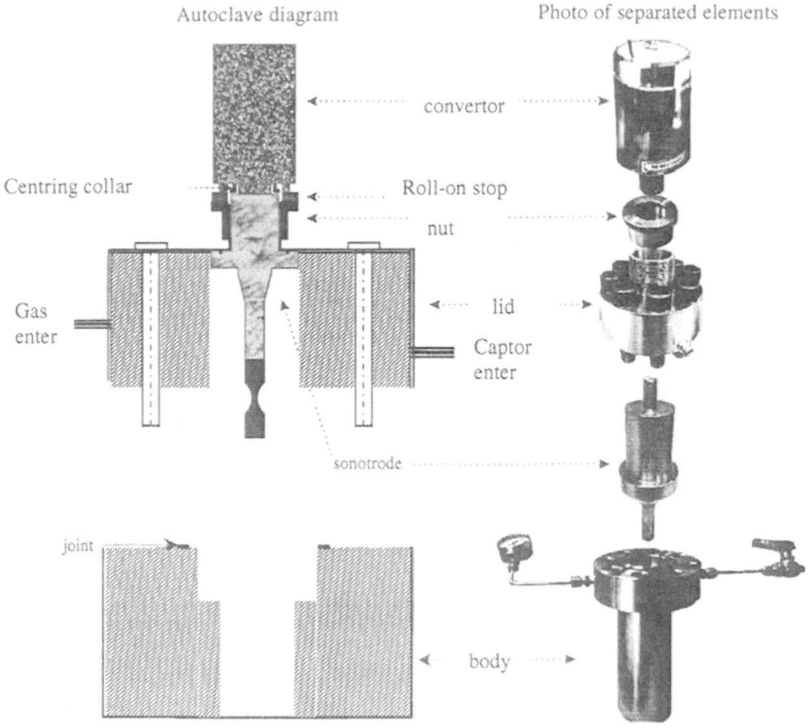

Figure 7 – *Autoclave description*

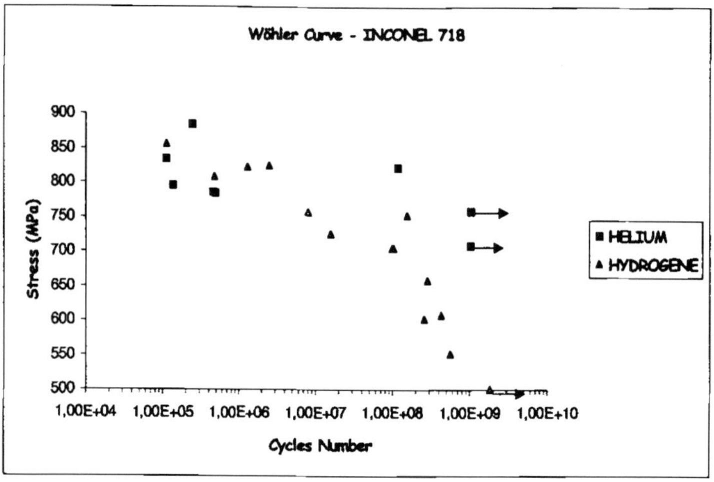

Figure 8 – Wöhler curve – INCONEL 718- R = -1

Ultrasonic Fretting Fatigue

Fretting fatigue is generally promoted by high frequency, low amplitude vibratory motions and commonly occurs in clamped joints and "shrunk-on" components. The surface damage produced by fretting can take the form of fretting wear or fretting fatigue where the materials' fatigue properties can be seriously degraded. Some practical examples of fretting fatigue failures are wheel shafts, steam and gas turbines, bolted plates wire ropes and springs. Fretting fatigue is a combination of fretting friction and fatigue process and involves in a number of factors, including magnitude and distribution of contact pressure, the amplitude of relative slip, friction forces, surface conditions, contact materials, cyclic frequency and environment. Great efforts have been made to quantify fretting fatigue in terms of these factors, but limited success has been achieved. More often, fretting fatigue characteristics are studied in the laboratory experimentally by using a contact pad clamped to a fatigue specimen in order to determine S-N curves, with and without fretting and thereby establish the fatigue strength reduction factor for a particular material. But these studies, generally performed on the conventional tension-compression fatigue machine at low frequency, have some inconveniences:

(1) The slip amplitude of fretting fatigue is usually coupled with the fatigue stress and to change the slip amplitude, pads with different gauge length are needed.

(2) The frequency is low and is not appropriate to simulate the small elastic vibration cycles at very high frequency of mechanical, acoustical or aerodynamical origin. In some industries, such as the automobiles and the railways, the determination of high cyclic fretting fatigue properties up to 10^8 or even 10^9cycles is necessary. This kind of experiment is time-consuming and uneconomic.

An ultrasonic fretting fatigue test technique at a frequency of 20 KHz has been developed, in which fretting slip amplitude can be changed individually without changing the fretting pads. Experiments were performed on a high strength steel and the results were analysed.

The fretting pad has also a cylindrical gauge profile. It is of the same materials as the specimen. The pads are held on the two sides of the specimen by two springs.

Figure 9 shows a schematic diagram of an experimental set-up. It consists of two parts. The first is the ultrasonic fatigue test machine, which has been widely used in fatigue tests for both endurance and crack propagation. Each element at the machine is designed to have a resonant frequency of about 20 kHz and an automatic unit maintains the whole system operating at the resonant frequency. The second part is a fixture to hold the two cylinder pads pressed onto the specimen by two springs. The normal contact force was measured and controlled by the displacement of the springs. Moreover, the use of the springs means that there will be a negligible fall off in load should wear occur. The axial loading experimental system was controlled by a PC.

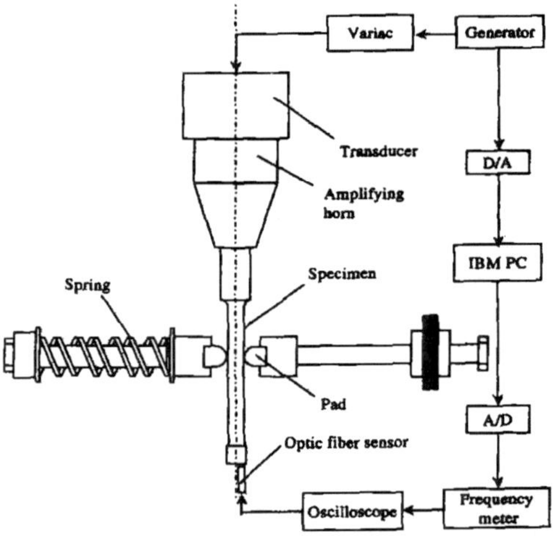

Figure 9 – *Schematic experimental system for ultrasonic fretting-fatigue*

The specimen for ultrasonic fretting fatigue has a cylindrical profile with different section and is asymmetrical to amplify the fatigue stress in the gauge length (see the distribution of the vibration displacement and stress in Fig.10. The specific length L is determined according to the need for the specimen to have a first longitudinal vibration resonant frequency of 20 kHz:

$$L = X_1 + X_2 = \{arctg[ctg(kL_1)S/S_1] + arctg[ctg(kL_2)S/S_2]\}/k$$

where k is a material constant, $k = 2\pi f\sqrt{\rho/E_d}$, S is the section area of the cylinder.

During the test, a maximum displacement is achieved at the free ends while the maximum strain (stress) is obtained in the center of the gauge length of the specimen (Fig. 10). In this test system, the fretting slip amplitude and the fatigue stress are the vibration displacement and vibration stress respectively at the point on the gauge length of the specimen where the pads are placed. They depend upon the position of the pad and the maximum vibration amplitude of the specimen. The latter is determined by the power of the generator and the amplification of the horn. In our experiment, this varies from 3 to 95 μm. By regulating the position of the pads along the specimen and by changing the power of the generator, either the slip amplitude or the stress of fatigue or both could be changed. As a result, these two parameters are decoupled.

Before the test, specimens and pads were carefully polished with emery paper. The pads were placed to the position of the specimen according to the slip amplitude and fatigue stress desired. After each test, the position of the pads was measured again, and the slip amplitude and fatigue stress were recalculated. During the test, the specimen was cooled by compressed air to decrease the temperature rise caused by friction and by the absorption of the ultrasonic energy. The normal contact force is 30 N and the slip amplitude is about 17μm.

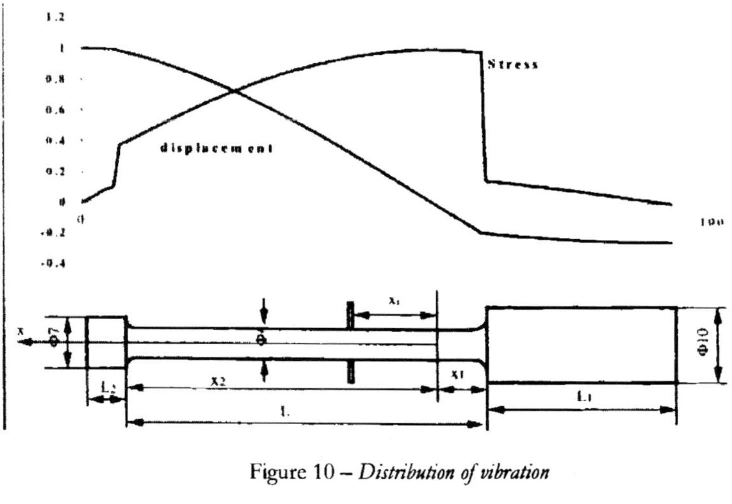

Figure 10 – *Distribution of vibration*

The conventional method of understanding the important variables which can affect fretting fatigue has been to generate *S-N* curves with and without fretting, allowing fretting fatigue strength reduction factors to be evaluated. Such a curve is given in Fig.11, which reveals that fatigue strength is significantly reduced by fretting fatigue, and the factor of reduction is of the order of 3 but varies with the number of cycles in a linear relation in the logarithm (Fig.12).

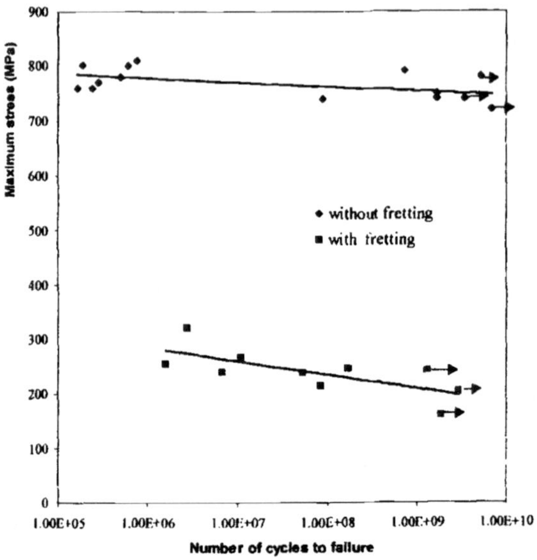

Figure 11 – *Fretting fatigue S-N Curve*

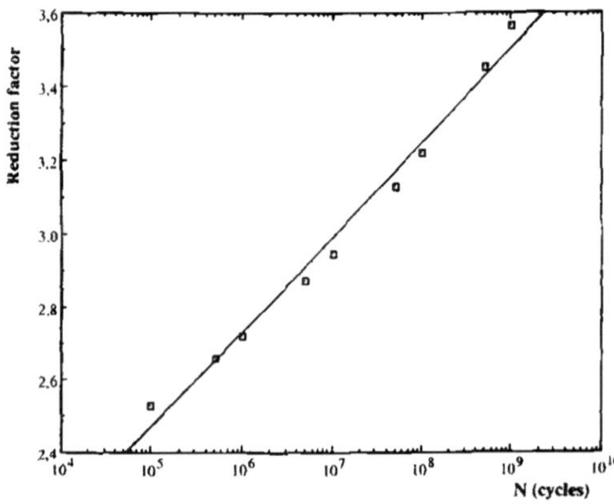

Figure 12 – *Fretting fatigue life reduction factor*

The experimental results in Figure 11 show that fatigue failure can occur after more than 10^7 cycles, and even over 10^8 cycles, which reveals that for fatigue design engineering, fatigue limits usually determined at 10^7 cycles are not reasonable. In return, this phenomenon demonstrates the importance of the high frequency fatigue test technique.

Fretting not only accelerates crack initiation but also increases the rate of crack propagation. But there exists a threshold of stress intensity factor in fretting fatigue, below which fretting cracks do not propagate. In this case, fretting scars are in the form of large ellipses in this test and considerable fretting wear is encountered over the entire contact area, at the surface of both the specimen and the pad. The contact surface increases with the stress cyclic numbers. Red oxide debris is observed at the contact surface and the examination of fretting scars demonstrates some fine cracks at the surface but non-propagation.

Conclusion

Special devices have been design to work in severe environments using a piezoelectric fatigue machine at high frequency (20kHz). Several advantages have been underlined.

- This new method is recommended to study the gigacycle fatigue regime of metals.
- The piezoelectric fatigue machine is able to operate at high temperature, cryogenic temperature, high pressure and fretting.
- The duration of a test is at least 400 times shorter than with a conventional machine. Thus this method saves considerable time and money.

References

[1] Mason, W.P. *Piezoelectronic Crystals and their Application in Ultrasonics,* Van Nostrand, New York, 1950, p. 161.

[2] Wu, T.Y., Ni, J.G. and Bathias, C. "Automatic in Ultrasonic Fatigue Machine to Study Low CrackGrowth at Room and High Temperature." ASTM STP, 1231, 1993, pp. 598 – 607.

[3] Bathias, C. and Ni, J.G. "Determination of fatigue limit between 10 and 10 cycles using an ultrasonic fatigue device", ASTM STP, 1993, 1211, pp. 151-152, 1993.

[4] Bathias, C. "Relation Between Endurance limits and Thresholds in the Field of Gigacycle Fatigue", ASTM STP 1372, 2000, pp. 135-154.

Mark L. Renauld,[1] Jonathan A. Scott,[1] Leroy H. Favrow,[1] Michael A. McGaw,[2] Michael D. Marotta,[1] and David M. Nissley[1]

An Automated Facility for Advanced Testing of Materials

Reference: Renauld, M. L., Scott, J. A., Favrow, L. H., McGaw, M. A., Marotta, M. D., and Nissley, D. M., **"An Automated Facility for Advanced Testing of Materials,"** *Applications of Automation Technology in Fatigue and Fracture Testing and Analysis,* ASTM STP 1411, A. A. Braun, P. C. McKeighan, A. M. Nicolson, and R. D. Lohr, Eds., American Society for Testing and Materials, West Conshohocken, PA, 2002.

Abstract: A novel facility has been developed for elastic-plastic-viscoplastic evaluation of structural materials through the integration of universal servo-hydraulic actuators connected to a standard servo controller and linked to an advisory control system automation package. Significant flexibility in terms of control mode, loading rates and end levels is achieved using the developed software/hardware interfaces. This technology enables complex waveform test profiles while ensuring machine tractability through closed loop control via feedback signals. Functions such as reloading, dwell, loading rate revision and mode switching can be programmed to trigger at any time during the test profile. Actions may target either axial or torsional actuator response since each control channel is fully independent. Signal conditioning and noise reduction for digital data acquisition are accomplished with "onboard" active filtering located on control system equipment as well as by employing post feedback active filtering provided by a commercially available digital dual-channel programmable filter system.

Keywords: advanced testing, cyclic loading, TESTExpress®, constitutive modeling, data reduction, MTS servohydraulic test system

[1] Technology Manager, Materials Engineer, Senior Engineer, Senior Engineer, and FAA/Industry Coordination Manager respectively, Pratt and Whitney, 400 Main Street, M/S 162-20, M/S 114-44, M/S 144-44, M/S163-07, M/S 163-07 East Hartford, CT 06108

[2] Owner, McGaw Technology, Inc., P.O. Box 26268, Fairview Park, OH 44126

Introduction

In 1998 Pratt and Whitney commenced an initiative aimed at enhancing all aspects of commercial and military fatigue life prediction methodologies. Central to the initiative resides an advanced material deformation model or constitutive model (Figure 1) with the capability of predicting elastic and inelastic material behavior. Model development tasks include both experimental and analytical efforts, with the experimental portion requiring advanced testing capability due to the unique data requirements needed for simplified constitutive models [*1*, *2*].

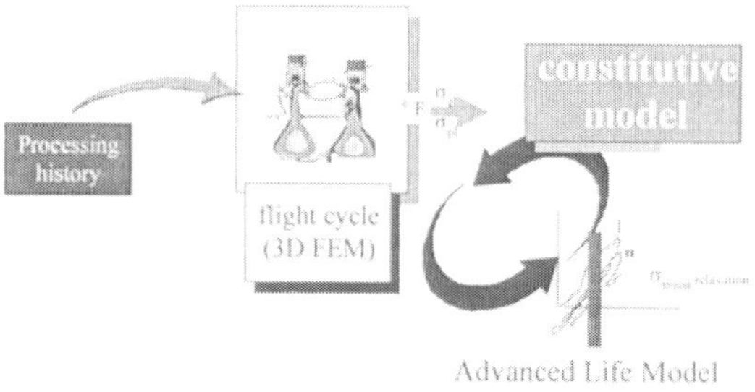

Figure 1 - *Implementation of constitutive model.*

The aggressive constitutive model development schedule, in conjunction with economic constraints, dictate the use of complex strain controlled profiles to capture strain rate dependent inelastic behavior that may be operative at a given temperature using combinations of strain end levels, strain rates and strain dwells. Recent advances in electronic and computer technology have enabled significant gains in experimental testing from machine control flexibility to all forms of data manipulation including data generation, collection, transfer, and reduction [*3*]. Complex strain-controlled experimentation requires state-of-the-art mechanical testing capability combining the ability to perform the tests while electronically acquiring data for rapid data reduction and evaluation. Additionally, as a unique material response is observed, the test facility must possess efficient flexibility to incorporate test profile modifications. To this end, the automated facility shown in Figure 2 was developed at United Technologies Research Center (UTRC) and has been on line for two years. The fully automated testing facility is comprised of universal servo hydraulic test equipment, having axial and torsional capability, coupled with an add-on automation package. This testing facility enables personnel to accommodate complex test protocols whether in concert with ASTM standards or addressing specialized testing. Further, this technology furnishes precise, accurate digital data supporting a relatively straightforward data reduction process.

Figure 2 - *Computer controlled servo-hydraulic test rig at United Technologies Research Center (UTRC).*

Component Description and Integration

An overall schematic of the "advisory control system" (MTS 458.20 MicroConsole coupled with McGaw Technology Inc, automation package) with a uniaxial load frame and Frequency Devices™ Filtering network are shown in Figure 3. The MTS MicroConsole alone is referred to as the "closed-loop servo controller" and the McGaw Technology Inc. automation package alone will be referred to as the "supervisory control system." Although the actual test system includes a tension-torsion load frame, a simple axial frame is shown for clarity. Torsional control is accomplished in the same manner as the uniaxial outlined procedure. Each key component will be discussed in greater detail.

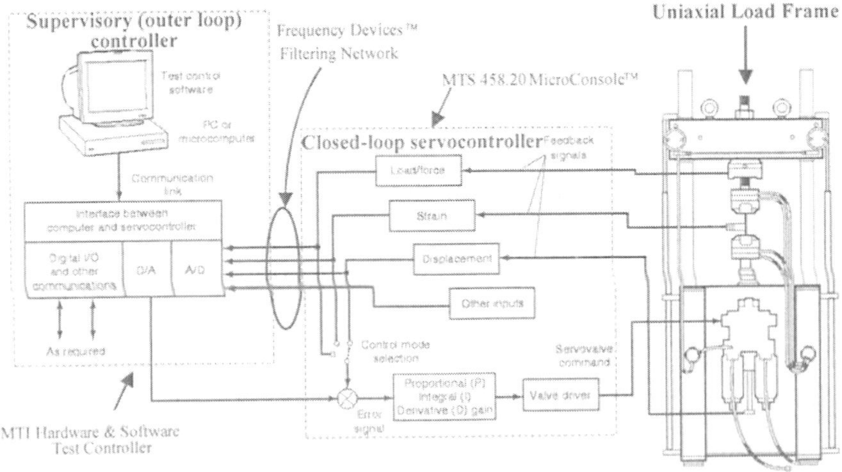

Figure 3 - *Component integration and interfacing for the "Advisory Control System" [4].*

Supervisory (outer loop) Controller Hardware and Software

At the heart of the facility is the "supervisory controller", with precursors described in references [*5*] and [*6*], that includes a rack mount hardware chassis "test controller" and components coupled with a PC driven software bundle. The use of the supervisory controller together with the closed-loop servo controller is often referred to as an "advisory control system." Such systems have a number of advantages over fully digital systems; analog servo controllers typically have a wider control bandwidth and the best examples of analog control technology feature low noise signal conditioning. Analog servo controllers afford a great deal of flexibility with regard to signal inputs and outputs enabling unusual or unique test requirements to be easily addressed (when combined with appropriate out board digital interfaces and associated software). Figure 4 shows the front view of the rack mount chassis, which provides the interface between the PC driven software and the closed loop servocontroller. This design provides independence from the PC during test execution. As a result, overall system durability is enhanced because the chassis is a skeletal system containing fewer applications running fewer components and is not affected by PC malfunction. This configuration provides additional flexibility whereby the PC is available for other activities such as data reduction from previous tests.

Figure 4 – *Supervisory rack mount chassis beneath analog recorder.*

All programming pertaining to test profiles and data acquisition is performed through the supervisory controller software "Workbench" control tree format. A distinct advantage of this system is the ease of programming based upon three fundamental routines: blocks, limits, and waveforms. Specific test parameters are established within the control tree utilizing user-friendly menus. Block routines set the number of iterations, algorithms (i.e. amplitude control during high frequency tests), cycle segments to acquire data and control mode. Limit routines support control mode switching, time-based limits, and reference limits such as load, strain, or stroke level crossing. Waveform routines set data acquisition parameters such as rates, channels on which to acquire data, and peak/valley specifications as well as the waveform itself. These waveforms may be full cycle commands such as sinusoidal or triangular, or may also be "singular" commands such as a ramp or hold. Test parameter modifications within a control tree are accomplished with relative ease satisfying the requirement for efficient execution of requested profile changes, as described above. A control tree consisting of blocks, limits, and waveforms is illustrated in Figure 5, as would be observed on the PC monitor. In this example, a simple ramp is followed by sequential sets of cycles with various strain excursions eventually terminated by a load reference limit of zero. A constant data acquisition rate was specified.

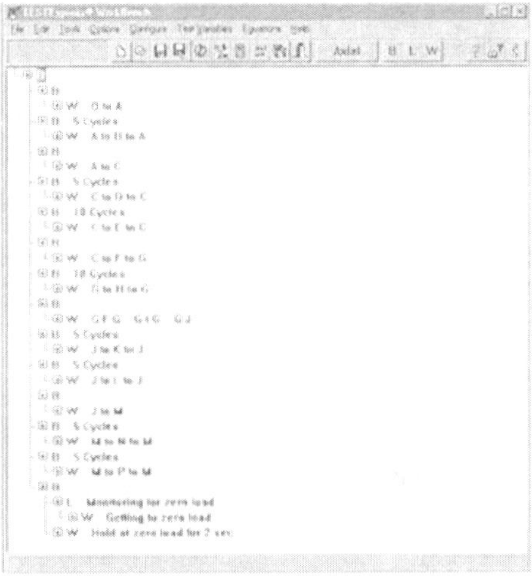

Figure 5 - *Control tree for complex strain controlled test profile.*

Standard data acquisition profiles are set up in order to optimize the number of data points, maintaining appropriately-timed high frequency sampling needed to fully capture material behavior while limiting file size. For example an extended dwell requires a slower time based data acquisition rate. Peak/valley data used to verify command end levels are achieved during cyclic profiles and to generate complete deformation curves, in conjunction with continuously acquired data, providing higher resolution than is typically recorded using analog systems such as X-Y recorders.

After test completion, test engineers may save the raw data in standard formats, like ASCII, for simple exchange with common spreadsheet packages. Templates may be created in the spreadsheets to convert load to stress or extensometer deflection to strain, based on specimen cross-sectional area and extensometer gage length, respectively. This reduced data, including graphic representations, may then be electronically distributed. The software embedded within the "Supervisory Controller" also allows significant internal data manipulation and evaluation. Intrinsic functions imbedded within the Workbench module permit the user to request properties such as Young's Modulus, 0.2% yield, and UTS, contingent on type of test performed. One can also utilize custom equations permitting a wide range of data analysis options an example of which would be optimized Ramberg-Osgood coefficients following acquisition of tensile test data.

Closed-Loop servocontroller.

A closed-loop servocontroller (Figure 6) relays test information from the "supervisory controller" to the load frame. The closed-loop servocontroller, in conventional closed-loop control fashion, dictates actuator motion via command signals to the servovalves in response to feedback from the load, strain or stroke transducers. Signal conditioners on board the closed-loop servocontroller, allow fine-tuning and system calibrations prior to each type of test. Fine-tuning system parameters affecting machine stability and accurate command/feedback signal matching include servo valve dither and PID (proportional, integral, and derivative feedback loop) gain settings [7]. Fine-tuning is performed as recommended under vendor-supplied manuals. An example of a calibration check is verifying extensometer full-scale range and linearity such that extensometer feedback is accurately converted to specimen strain. This can be accomplished by placing the extensometer on a commercially available calibrator and performing the required procedure ensuring correct calibration. This calibration check is performed more often than others, such as for the load cell, since specimen fracture is common under high strain testing conditions and extensometer rods are often broken or chipped. All calibrations are performed as frequently as recommended by the equipment manufacturer.

Figure 6 - *MTS 458.20 Axial/Torsional MicroConsole™*

Once the control tree has been created, a PC based software module is initiated and, upon user command, electronically transfers test instructions to the "supervisory controller." During test execution all combinations of feedback channels are available for graphical, scaled viewing on the PC display. Analog output is also provided for additional data acquisition methods – oscilloscopes, strip chart recorders, X-Y plotters, etc. Upon test completion, data is transferred automatically from the Rack Mount Chassis to the PC.

Filtering

Feedback signal filtering is employed to prevent potential digital data corruption due to erroneous signals induced by other equipment common to most lab atmospheres. RF heating, for instance, is generally recognized as a major noise contributor. To alleviate concerns of both constant and intermittent noise sources, a filtering system was introduced ensuring the majority of system and background noises are omitted from data collection (Figure 7). Typical feedback noise levels with filtering invoked are on the order of ten (10) to fifteen (15) millivolts peak to peak. This is consistent with values quoted by vendor specifications. Command signals are not filtered thereby preserving the integrity of the desired machine control. After numerous trials a cutoff frequency of 0.7Hz and a gain of one (1) db yielded the smoothest data without any artificial corruption brought on by the filtering system [*8*]. Having established these values, the post filtered signal was verified for fidelity, aliasing, etc.

Figure 7 - *Frequency Devices™ Model 9002 Digital Filter.*

Test Procedure

Program Validation and System Tuning

For the constitutive testing application, strain control is typically implemented due to the imposed large elastic and inelastic material strains. An extensometer calibration check is performed prior to each test since the quartz extensometer rods often break after each test and are replaced with new rods. All other transducers are calibrated and verified according to vendor specifications and intervals. For elevated temperature testing, a three zone, clamshell type furnace is used to minimize gage section thermal gradients. At room temperature, the initial extensometer rod separation is slightly reduced to account for thermal growth. After heating, temperature is maintained for 30 minutes and thermal strain stability is verified by the extensometer feedback. The actual extensometer gage length is then recorded for strain calculations during data reduction.

Test profile execution must be validated to ensure the rig performs as programmed. First, the control tree profile is executed without hydraulic pressure, to verify command signal endlevels, dwells, ramp rates, etc. Second, a dummy specimen of equal compliance to the actual test specimen (Figure 8) is inserted in the load train and is subjected to a trial run with hydraulics on. This data is transferred to a PC, command and feedback signals are compared and appropriate closed loop PID adjustments are made on the closed-loop servocontroller. Additionally, the command and feedback signals are magnified to reveal relative signal-to-noise ratios and to ensure previously established filtering parameters are acceptable.

Figure 8 – *Constitutive modeling test specimen.*

Data Reduction

Upon test completion, the rack mount chassis uploads two "result" data files to the PC, one simply serving as a back-up. The Report software module is executed to convert this data file to standard file types, such as ASCII text, thereby enabling transfer to software packages designed for data manipulation and handling. For example, Microsoft Excel can be used to convert measured load to stress, relative extensometer rod displacement to strain, plot the data, and interrogate material response. Once the digital data is reduced, a qualitative, and occasionally a quantitative, comparison is made to analog instrumentation traces, which are routinely obtained.

Application and Results

Constitutive Profile Design

For the constitutive modeling effort, a set of standard profiles with unique sets of axial strain endlevels, strain rates and dwells are designed for low (rate-independent), intermediate and elevated (rate-dependent) temperature evaluation. Fewer than 30 specimens are targeted for full alloy characterization, using approximately 20 samples for model calibration at specified temperature intervals from room temperature to the alloy's solution temperature. Additional specimens baseline general inelastic characteristics and provide evaluation and verification data using strain sequences or temperatures other than the standard profiles from which model constants are regressed. Each test is conducted with a new specimen under uniaxial isothermal conditions. In addition to meeting all technical requirements for material characterization, the profiles are designed for control tree creation and evaluation, machine setup and calibration, test completion and data reduction within a standard 8-hour workday. In fact, some tests require more time for temperature elevation and stabilization than all other aspects of the job combined.

A profile schematic, which formed the basis for the control tree presented in Figure 5, is shown in Figure 9 with key points represented by letters. This type of template matches the control tree concept in that standard routines are established with simple strain/stress end level adjustment within Workbench module. Generally, one strain point (letter A) is established on an absolute scale, say proportional limit or 0.2% offset yield strength, with other strain points determined relative to the fixed strain value and lettered B-P. Using this approach, the control tree can be used for a given material at multiple temperatures or different materials possessing various elastic and inelastic properties. Experimental data from a nickel base superalloy tested using the Figure 5 control tree is presented in Figure 10. A significantly different profile, with additional complexity, used on the same material at an elevated temperature is shown in Figure 11. As a side note, excellent correlation can be observed between the experimental (solid) and model (dashed) traces in Figures 10 and 11. Multiaxial (tension-torsion) profiles have been created and facility modifications are nearly complete at the writing of this paper.

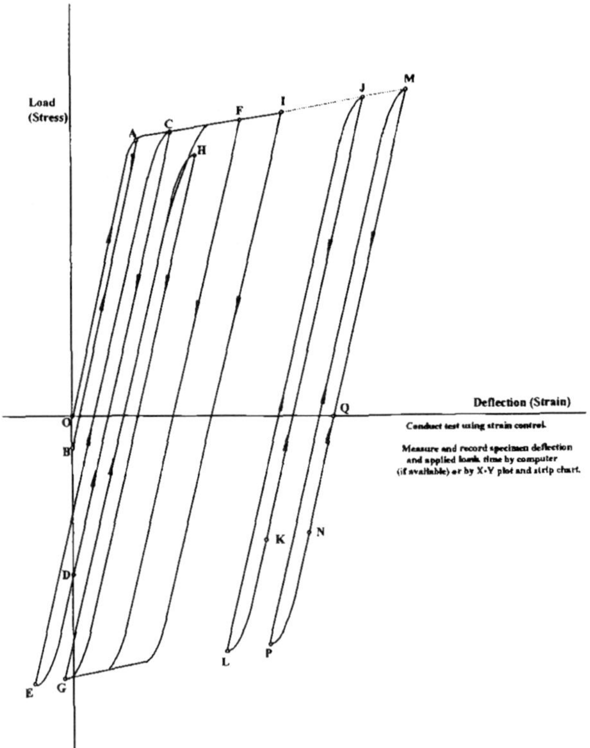

Figure 9 – *Schematic representation of strain controlled profile.*

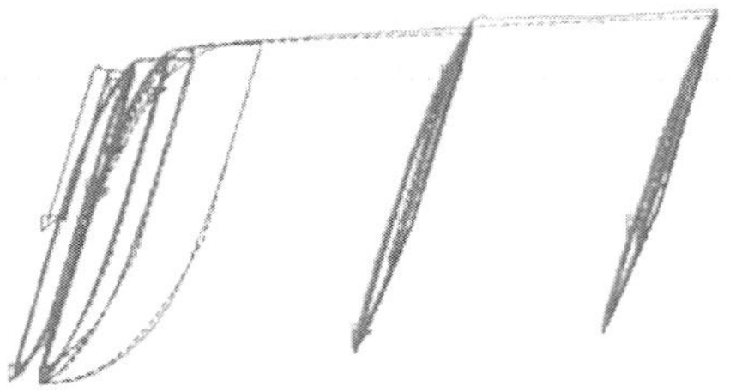

Figure 10 – *Room temperature experimental data on a nickel base superalloy (solid line) and constitutive model correlation (dashed line).*

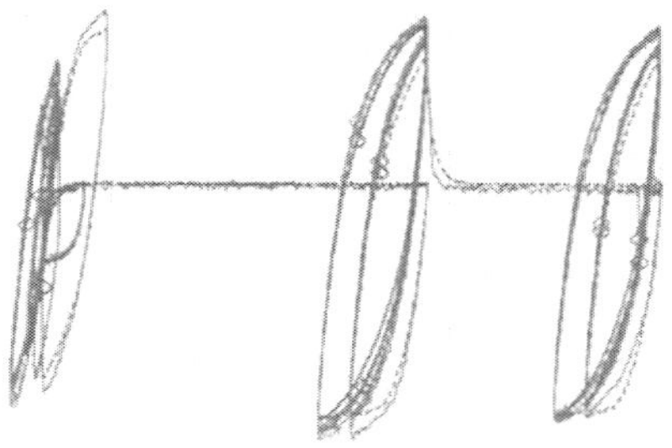

Figure 11 – *Elevated temperature experimental data on a nickel base superalloy (solid line) and constitutive model correlation (dashed line).*

Conclusions

Comprehensive, cost effective material testing has been achieved through the integration of commercially available software and hardware providing state-of-the-art experimental capability. The system consists of components from MTS, MTI, and Frequency Devices™ and permits user friendly creation and modification of control trees enabling variable combinations of mode control, dwells and user intervention algorithms. This allows simulated component service cycles, which now generates data from a single test previously requiring multiple tests. Significant data acquisition capabilities are available with outputs in standard formats (ASCII) for transfer to other software. This facility has exhibited exceptional reliability under nearly constant operation for two years. The machine capabilities are currently being expanded to include the torsion control channels (i.e. actuator rotation, rotational strain, and torque).

Acknowledgments

Equipment acquisition has been supported under United Technologies Corporate funding while developmental and operational activities have received Pratt and Whitney IR&D sponsorship managed by Dave Szafir and Richard Holmes. The authors also appreciate the component integration assistance provided by Ron Holland and Gene Roman of UTRC and Kevin Rogers and Dave Taus of MTS.

References

[1] Renauld, M. L, Annigeri, R. and Zamrik, S. Y., "Viscoplastic Model for Thermomechanical Fatigue," *Low Cycle Fatigue and Elasto-Plastic Behavior of Materials*, K-T. Rie and P. D. Protella, Eds., Elsevier Science Ltd, 1998, pp. 155-160.

[2] Nissley, D. M., Meyer, T. G. and Walker, K. P., "Life Prediction and Constitutive Models for Engine Hot Section Anisotropic Materials Program," NASA CR 189222, August, 1992.

[3] McGaw, M. A., Materials Testing Software LEW-16160, COSMIC, 1995.

[4] Halford, G. R., Lerch, B. A. and McGaw, M. A.: Fatigue, Creep Fatigue, and Thermomechanical Fatigue Life Testing, ASM Handbook, Volume 8, Mechanical Testing and Evaluation, pp. 686-716. Howard Kuhn, Dana Medlin, Eds. ASM International, Materials Park, Ohio, 2000. ISBN 0-87170-389-0

[5] McGaw, M. A., Bonacuse, P. J., "Automation Software for a Materials Testing Laboratory", ASTM STP 1092, *Applications of Automation Technology to Fatigue and Fracture Testing*, A. A. Braun et al., Eds., American Society for Testing and Materials, 1990, pp. 211-231.

[6] TESTExpress® Reference Manual, McGaw Technology, Inc., 1998.

[7] Operation Manual for MTS Test Systems Containing a 458.20 MicroConsole™, 1992.

[8] Frequency Devices™ Model 9002 Operator Manual, July, 1989.

Bonnie R. Antoun[1] and Louis F. Coffin, Jr.[2]

Experimental Technique for Monitoring Fatigue Crack Growth Mechanisms During Thermomechanical Cycling

---

Reference: Antoun, B. R. and Coffin, L. F., Jr., "**Experimental Technique for Monitoring Fatigue Crack Growth Mechanisms During Thermomechanical Cycling,**" *Applications of Automation Technology in Fatigue and Fracture Testing and Analysis: Fourth Volume, ASTM STP 1411*, A. A. Braun, P. C. McKeighan, A. M. Nicolson, and R. D. Lohr, Eds., American Society for Testing and Materials, West Conshohocken, PA, 2002.

Abstract: A fully automated thermomechanical fatigue test system capable of extremely sensitive crack growth measurements was developed. This system was used to conduct thermomechanical fatigue tests on titanium matrix composites to study the crack growth behavior throughout cyclic testing as well as during individual loading cycles. Experiments were performed in bending on specimens with a realistic initial defect, a corner crack geometry, for composite materials. Data was collected during the experiments using the reversing dc electrical potential method and changes in crack dimensions were determined via an inverse solution to the electrical potential field. Isothermal, in-phase and out-of-phase thermomechanical fatigue crack growth experiments were conducted with test temperatures ranging from 204°C to 538°C.

Keywords: thermomechanical, fatigue (materials), crack propagation, crack closure, corner crack, composite, titanium matrix composite, electrical potential

Thermomechanical fatigue (TMF) experiments are much more difficult, expensive and time-consuming to conduct than isothermal fatigue experiments. However, isothermal fatigue tests do not capture many of the important damage mechanisms that occur during varying temperature conditions. During thermomechanical cycling, the alternating activation of high and low temperature mechanisms results in a unique combination of effects that may be more detrimental than any of these mechanisms could produce isothermally. Due to the internal structure of the composite itself, TMF is one of the most common and severe in-service loadings. Unlike TMF in monolithic materials which requires a thermal gradient, a thermal shock or an external constraint during

---

[1]Senior member of Technical Staff, Materials Mechanics Department, MS 9042, Sandia National Laboratories, P.O. Box 969, Livermore, CA, 94551.
[2]Distinguished research professor (Retired), Department of Mechanical Engineering, Aeronautical Engineering and Mechanics, Rensselaer Polytechnic Institute, 15th Street, Troy, NY, 12180.

temperature changes, TMF in composites can occur merely by a change in temperature. This is due to the coefficient of thermal expansion mismatch between the fibers and the matrix. During temperature changes, the fibers act as an internal constraint, thereby producing thermal stresses in both the fiber and the matrix. These thermal stresses are superimposed on any applied mechanical cycling, resulting in TMF of the composite.

Studies of the TMF behavior of titanium matrix composites have shown that TMF life can be considerably shorter than isothermal fatigue life, depending on the thermomechanical phasing [*1-6*]. It has also been observed that during TMF, cracks initiate very early in life, resulting in the composite fatigue life being determined by crack growth [*7,8*]. Cracks tend to initiate at the specimen edge and propagate as corner cracks. The main characteristic of fatigue cracks in titanium matrix composites is the bridging effect of the fibers left in the wake of the crack, which slows the crack growth rate substantially compared to the matrix material alone. Based on these observations, it was determined that the most critically needed study was of crack growth during TMF. It was also apparent that using a specimen geometry that closely replicates naturally occurring damage would be extremely useful. This paper describes the study completed on the crack growth from initial corner cracks during TMF, with emphasis on the test system and experimental techniques that were developed.

Experimental Apparatus

A TMF crack growth test system was required in which specimens could be subjected to a loading environment that simulates the mechanical and thermal conditions that the metal matrix composite would be exposed to during service. A bending test specimen was chosen over a tension test specimen for the following reasons: fatigue loading by bending allows tension-compression fatigue testing, loading in bending provides a unique opportunity to conduct two experiments on the same specimen, loading in bending is a better simulation of thermal and stress gradients experienced in-service, and the availability and cost savings in using a bending fatigue test system over a servohydraulic tension-tension test system. The major disadvantage of using a bending test specimen is the added complexity of all of the analyses and interpretations, because the crack is growing in two dimensions in a varying stress field.

The major design considerations for the TMF crack growth test system were: (1) maximum thermal cycle range from 150°C to 650°C, (2) 16-ply composite specimens, (3) fatigue loading by bending, (4) variable loading rates and waveform types, and (5) accurate measurement of crack growth. The design was also influenced by the necessity to build a low cost test system.

All three components of the TMF crack growth test system are computer-controlled. Software was developed to perform all control and data acquisition tasks necessary for this fully automated, real-time control, closed loop system. Two files are used for storage of the data collected in each experiment: the hold time data file is used to store data collected during each hold period at maximum and minimum stress for the experiment duration, and the individual cycle data file is used to store data collected throughout entire individual loading cycles of interest. A significant amount of work is required to reduce and process these data files for crack dimension predictions and is described in Ref. *9*.

Mechanical Loading System

A system was developed to apply mechanical loading to a cantilever beam specimen in bending by driving an eccentric cam with a motor as shown schematically in Figure 1. It is based on a previous system [*10*], modified for high temperature and controlled loading waveforms. A Compumotor S57-51 stepper motor with a Bayside 20:1 precision gearhead was chosen to meet the torque and speed requirements. Together they provide excellent resolution, 500 000 steps/revolution, for precise control. A Compumotor SX6 motor controller is used to control the stepper motor. Commands are supplied to the controller by the testing software via one of the computer serial ports using the RS 232 protocol. To measure the applied strain, two fatigue strain gages are connected in a Wheatstone bridge circuit. The bridge output is connected to the HP 3457A digital voltmeter and read by the testing software via a GPIB card in the computer using GPIB protocol. An inductive proximity switch that senses the passing of a metal target located on the drive wheel is connected to an electronic counter to display the applied cycles. This provides visual information in the laboratory but is not connected to the computer since cycling is controlled and monitored by the computer and motor controller.

Allowing the loading system to run continuously through complete revolutions would result in a sinusoidal type loading waveform. However, variable loading and unloading rates, waveform types and hold times were needed to perform the TMF experiments, as was the ability to synchronize the applied loading with the temperature cycles. A method was developed to produce controlled loading waveforms with this system by continuously varying the angular velocity of the motor and changing the rotation direction after each half cycle. The waveform, in the form of beam deflection as a function of time, is prescribed. This is used as input into a program that was written to perform a numerical integration of the kinematic velocity relationship of the rotating mechanism of the beam bending machine. At 1/8 second intervals, the program calculates the angular velocity required to produce the prescribed waveform. The output of the program is two data files: 1) beam deflection and 2) angular velocity at incremental time periods through a single load waveform. These two data files are read as input into the testing software. A table lookup scheme is used to access these files during each cycle of the experiment. The angular velocity is used to calculate and update the motor velocity and the beam deflection is used to calculate and control the experiment temperature, thereby synchronizing the strain and temperature in each cycle. The deflection waveform and the angular velocity for experiment 133-5 are shown in Figure 2. The resulting strain waveform for a cycle during the experiment is shown in Figure 3.

Temperature Control System

The components of the temperature control system were designed based on the heat transfer analyses described in Ref. 9. The cooling portion was found to be the time limiting part of the thermal cycle. Two-dimensional finite element heat transfer analyses were performed to determine cooling rates, temperature profiles, and temperature variation through the composite specimen thickness for various thermal cycle limits. The measured temperature profiles were found to match very well with those predicted using

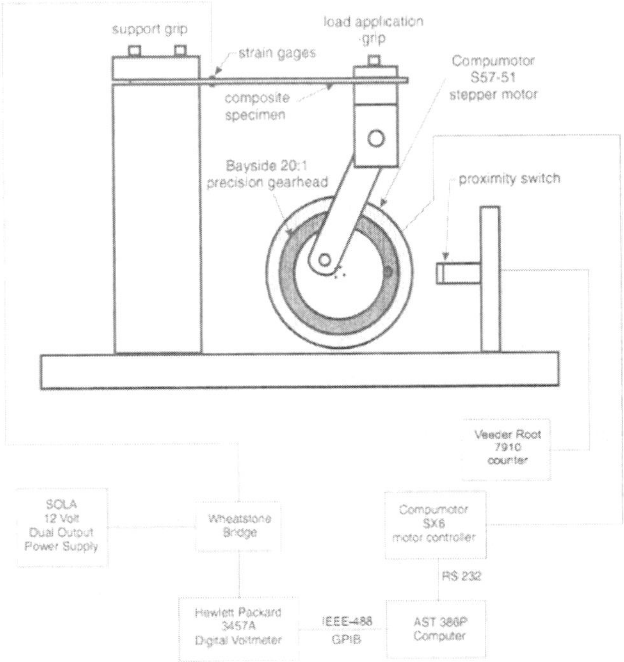

Figure 1 - *Mechanical loading system.*

finite element analyses. Based on the analyses, it was determined that for the most extreme thermal cycle, 538°C to 204°C, cooling times less than 120 seconds produced temperature variations through the thickness that were not negligible.

A schematic of the temperature control system is shown in Figure 4. The specimen is heated with a hand wound resistance furnace of about 70 Ω. Power is supplied to the furnace directly by a Eurotherm 94C temperature controller. A type K thermocouple, attached to the top, center of the specimen at the crack location, is used for feedback. Commands are sent to the temperature controller by the testing software via one of the computer serial ports using the RS 422 protocol. A separate type K thermocouple is attached at the same location, only on the bottom of the specimen, for independent monitoring of temperature. A third, type T thermocouple is used to monitor temperature at the location of the strain gage. The output from both thermocouple adapters are connected to the HP 3457A digital voltmeter and read by the testing software using the GPIB card in the computer. The grips on either end of the specimen are cooled by a continuous, fresh supply of water. Copper tubing was soldered to the brass plate and silver soldered to the steel plate.

Optimization of the temperature control parameters allowed very good control with minimal cycle-to-cycle variations. During the thermomechanical experiments, the measured temperature waveform differed from the programmed control waveform by two percent or less. A typical response during experiment 133-5 is shown in Figure 5.

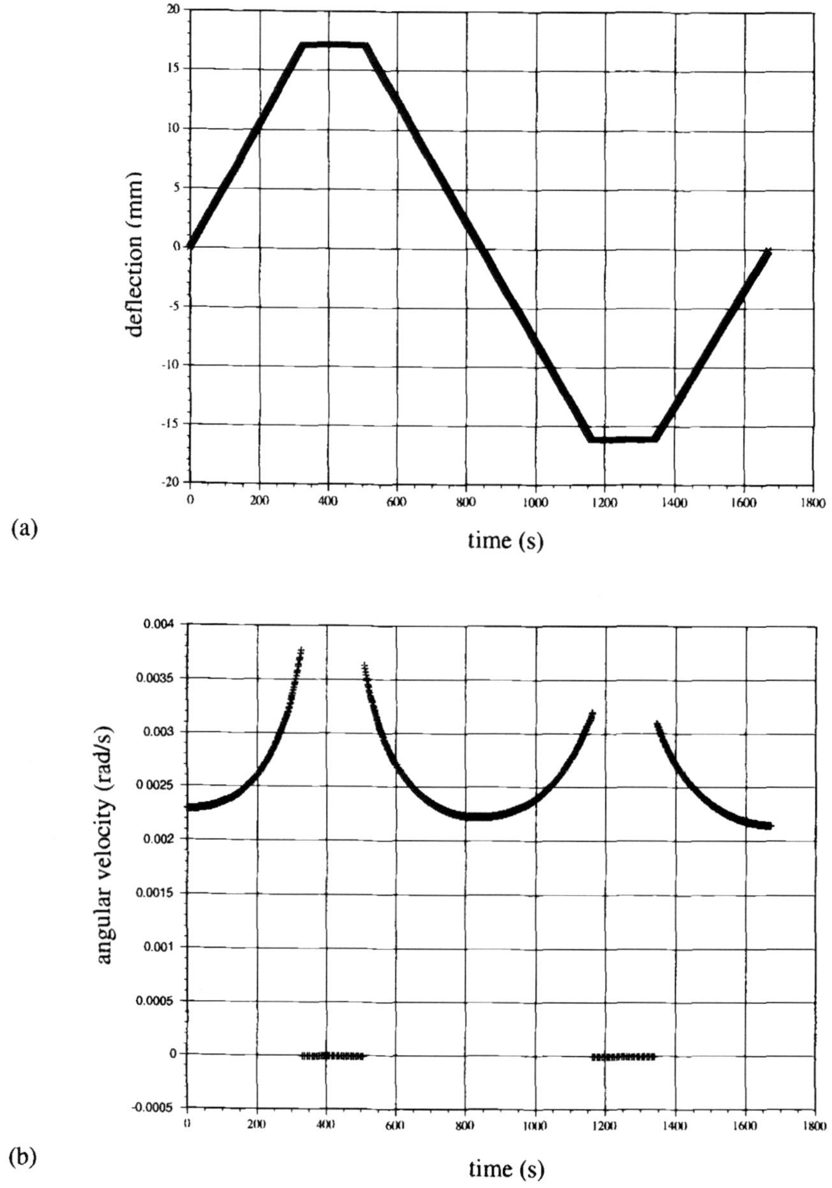

Figure 2 - *(a) Loading waveform and (b) Angular velocity through one loading cycle of experiment 133-5.*

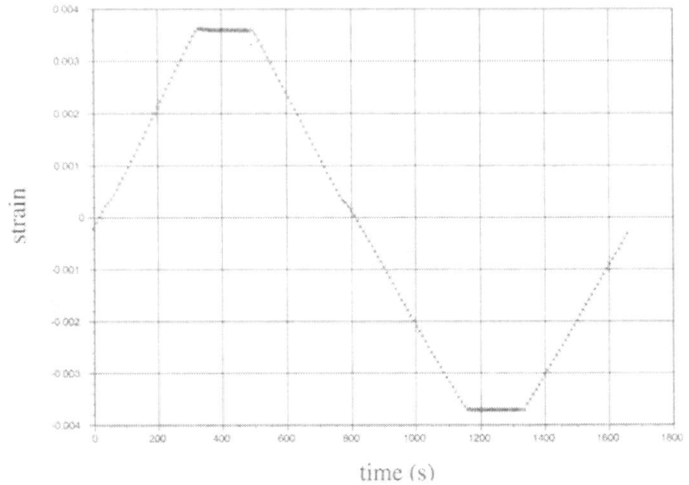

Figure 3 - *Measured strain during one cycle of experiment 133-5.*

Crack Dimension Measurement System

Data for monitoring the crack dimensions were measured using the reversing dc electrical potential method [*11*]. A schematic of the system is shown in Figure 6. A Sorensen SRL 40-12 dc power supply supplies a constant direct current of 12 A to the specimen as measured by a Fluke 90A current shunt. The signal from the current shunt is connected to one of the channels of the HP 3457A digital voltmeter. A solid state polarity reversing switch is controlled by logic levels of 0/5 V dc signaled by the testing software through the parallel port of the computer. This reverses the current every 2 seconds, producing a square current wave.

Six electrical potential probe pairs are attached to the specimen to measure crack growth, three probe pairs for each crack. The locations of the probe pairs were determined by performing numerical analyses to maximize sensitivity to crack growth over the range of crack dimensions expected [9]. Each probe pair is connected to a channel of the HP 3457A digital voltmeter. All input channels to the voltmeter are read by the testing software using GPIB commands sent through the GPIB card in the computer.

Specimen Preparation and Test Procedure

Material

The material used in this study was a titanium based metal matrix composite obtained from the Air Force Wright Laboratory. The matrix material is TIMETAL®21S (β-21S), a metastable β titanium alloy with the composition Ti-15Mo-3Al-2.7Nb-0.2Si in weight percent. The composite is reinforced with SM1240 unidirectional, continuous, 100 μm

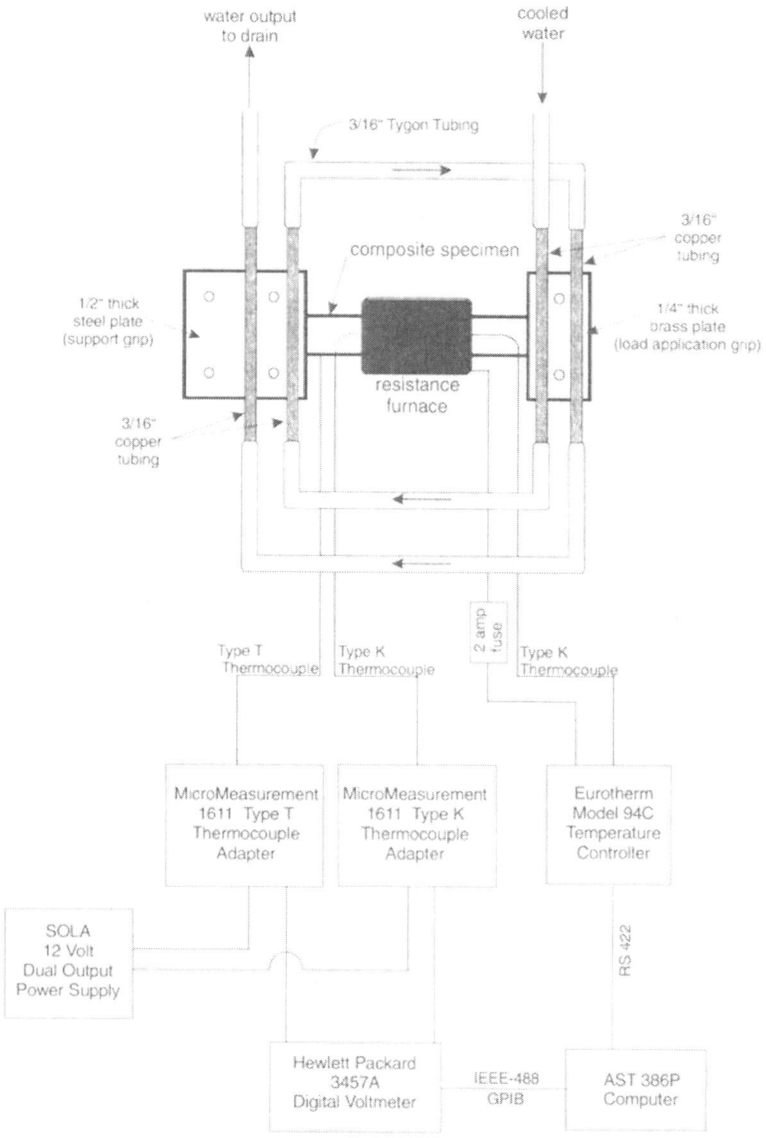

Figure 4 - *Temperature control system.*

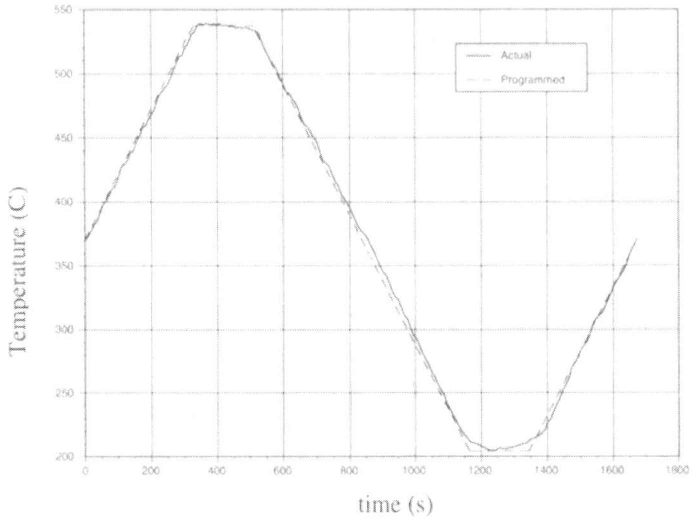

Figure 5 - *A comparison of measured and control temperature waveforms of experiment 133-5.*

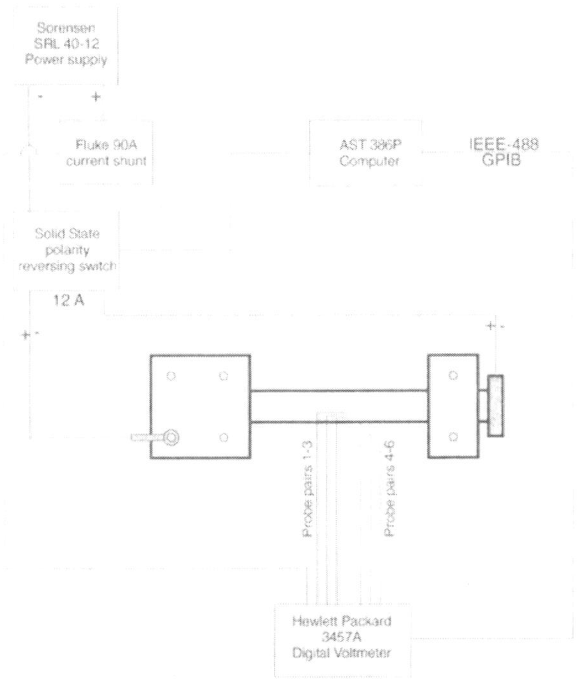

Figure 6 - *Crack dimension measurement system.*

diameter silicon carbide fibers. All of the Sigma/TIMETAL®21S composite specimens were machined from 16-ply thick plate number NIC-133, and were oriented such that the fibers were parallel to the specimen length, resulting in unidirectional 0° specimens, about 2.1 mm thick. The Department of Air Force machining guidelines were followed, which require diamond wheel machining with coolant and with a slow feed rate, 3/8 to 1 inch (9.5 to 25.4 mm) per minute. Machined edges are required to have an RMS 125 finish, with no burns, cracks, delaminations or inclusions. Specimens were cut to a nominal width of 15.24 mm (0.6 inches).

The next step was to machine a starter notch for each of the corner cracks in the specimen. It was necessary to have these notches very small, about 0.5 mm deep by 0.5 mm long, and sharp or as narrow as possible. A diamond wheel would have produced a notch too wide, and EDM was found to be unpredictable and unsatisfactory at this small scale. Therefore, a diamond wire, 0.002 inches (0.0508 mm) in diameter, was used to saw the notches manually under a light microscope until the notch depth and length dimensions were reached. This method was very satisfactory, providing sharp starter notches that produced consistent precracking results.

Strain Gage Installation

The strain gages used in this study were required to withstand high temperatures up to 150°C and a few thousand loading cycles. MicroMeasurement type WK-06-062ED-350 fatigue gages were selected. These gages have a small length, 6.2 mm, which minimizes the length over which strain is averaged. The coefficient of thermal expansion of the gage is matched closely to the specimen and the high 350Ω resistance allows high accuracy measurements. One strain gage was attached to each side of the specimen, at the same location along the length, 6.35 mm (0.25 inches) from the edge of the grip during testing, as shown in Figure 7. This location was chosen to produce a high strain signal within the fatigue limit of the gage. The two gages were wired in a half bridge, allowing double accuracy measurements that were not affected by temperature changes, due to cancellation in the half bridge configuration.

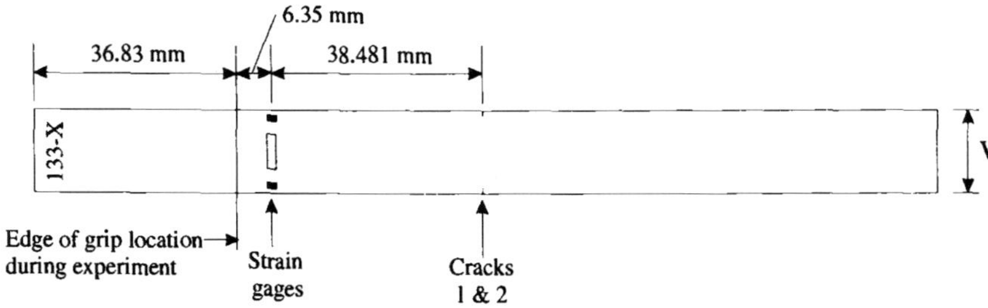

Figure 7 - *Location of the strain gages on the specimen.*

Precracking

Specimens were subjected to room temperature cycling to produce sharp precracks from the notches. This ensured a consistent starting point for crack growth measurements in the experiments. A software program was written and used to control the cycling for precracking. This program is similar to the thermomechanical testing software but tailored to precracking. During precracking, the specimen was installed in the test frame backwards, with the notches close to the edge of the fixed end grip and the strain gages towards the free end grip. The notches were checked periodically for crack growth under maximum load through a microscope. Each specimen was cycled until Cracks 1 and 2 were of equal length, typically 0.4 mm beyond the starter notch. The starting crack size correlated to an applied stress intensity factor range of about 25 Mpa$\sqrt{m}$ for isothermal tests, a lower and higher value respectively for in-phase and out-of-phase TMF, due to the thermal contribution to the applied stress range.

Electrical Potential Probe and Thermocouple Attachment

A set of six electrical potential probe pairs were prepared using 0.002 in (0.0508 mm) Chromel wire. Each probe wire was electrically shielded in high temperature sleeving. Probe pairs were made by twisting two shielded probes together to reduce noise in the electrical potential signals. The probe pairs were long enough to reach from the specimen directly to the test system and were reused for each experiment, after trimming the portion exposed to high temperature.

The probes were attached to the specimen by spot welding, which proved to be very successful as only a few probes fell off during the whole testing program and were easily reattached. The probe pairs were attached to straddle the two corner cracks on the specimen surface. The probe pairs were fastened to the specimen by tying them flat against the surface to minimize movement during testing.

During each experiment, three thermocouples were used for temperature control and monitoring. All were attached by spot welding. One type T thermocouple, used for moderate temperatures, was located 6.35 mm from the edge of the fixed end grip on the thickness face of the specimen to monitor temperature at the strain gage. Two type K thermocouples, used for high temperatures, were located at the center of the width of the specimen surface at the location of the cracks, one on each surface. One was connected to the temperature controller and the other was connected directly to a thermocouple adapter for monitoring.

Fatigue Crack Growth Experiment

The testing software TMFLITE [9], developed specifically for this research, was used to run the crack growth experiments. The experiments were conducted in strain control and were fully automated: the decision to end an experiment was based on the number of applied cycles and the estimated crack growth rate as well as the amount of increase observed in the electrical potential measurements. Specimen unloading and cooling was also software driven. Since the mechanical loading was applied in bending, the top and bottom surfaces of the specimen were subjected to equal but opposite stresses during cycling. Therefore, two fatigue crack growth experiments were conducted simultaneously on each specimen: two identical isothermal experiments for the isothermal conditions and two experiments 180° out of phase for the thermomechanical conditions, resulting in one in-phase and one out-of-phase TMF crack growth experiment.

Postcracking

At the completion of each experiment, the specimen was removed from the test system. Each of the probes was removed from the specimen under a microscope where a more accurate measurement of the probe spacings was made. Specimens were then reinstalled in the test system in the precracking configuration. The specimen was cycled at room temperature to grow the cracks beyond their final size reached in the experiment. This was done to mark the crack growth region clearly and to ensure that upon breaking the specimen the region would be visible for SEM analysis. A software program similar to the precracking software was used to control the cycling for postcracking. The notches were checked periodically under maximum load, with a microscope, for crack growth. The specimen was cycled until each of the cracks had extended at least 0.5 mm.

The specimens were cut in half lengthwise using a diamond saw, which separated cracks 1 and 2 into individual specimens. These specimens were installed into a three point bending load frame and loaded until failure. The cracks were then visible for viewing under the optical microscope and SEM.

Data Analysis

The measurement of crack growth in such a detailed manner as was desired in this study required both an immense amount of data and the proper analytical tools. Development of the electrical potential solution for a corner crack geometry as a function of probe location and specimen geometry, described in detail in Ref. *9*, was based on similar solutions developed at General Electric Corporate Research and Development Center [*12-16*].

The data collected during the experiments were stored into two types of data files: measurements collected during both hold times of each cycle and measurements collected throughout entire individual loading cycles. After considerable data reduction, a computer program was used to process the electrical potential data by a procedure that minimizes the least-squares error to give the best fit for the quarter-ellipse shaped crack dimensions, where a is the crack length and c is the crack depth [*9*]. For each set of probe data, the program calculates a, c, square root of the crack area and center offset of the crack. Individual loading cycle files of non-isothermal tests are subjected to an additional step of correcting for the thermal contribution to the electrical potential voltage. As an example, the normalized electrical potential for the three probes of test 133-5-1 is shown in Figure 8. Probe pair 1 is located closest to the edge of the specimen, followed by probe pairs 2 and 3. The predicted crack dimensions from these electrical measurements are also shown in Figure 8.

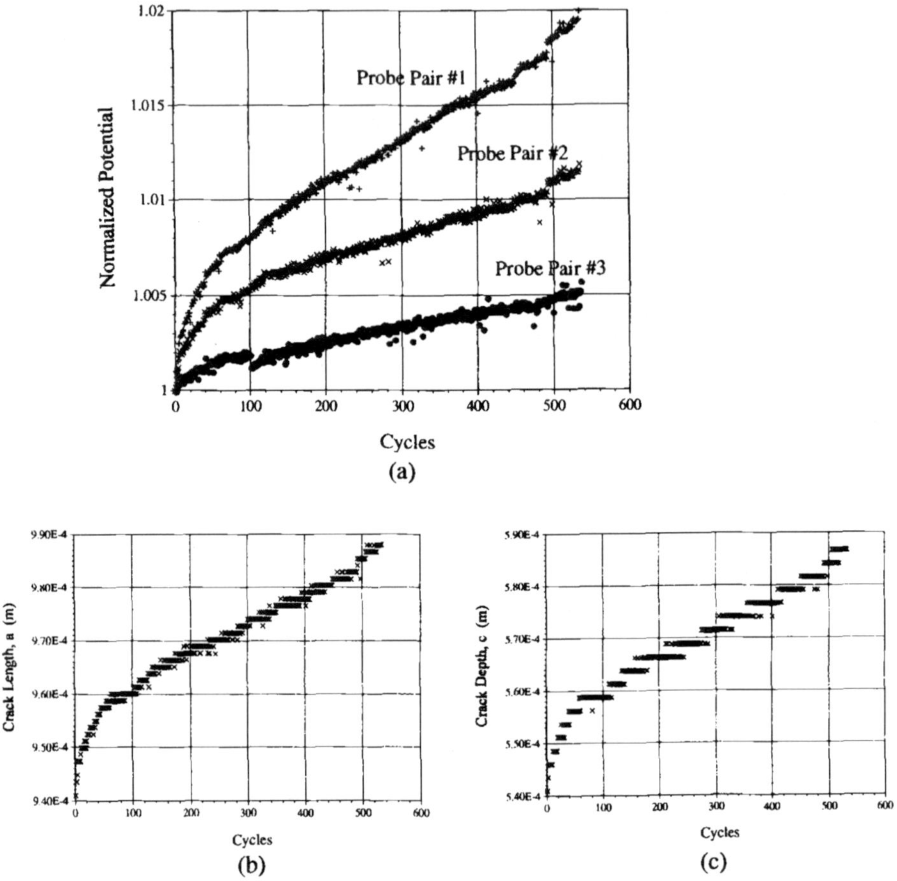

Figure 8 - *Test 133-5-1 (a) normalized electrical potential data, (b) crack length prediction, (c) crack depth prediction.*

Crack growth in the composite is governed by the stress intensity factor range at the crack tip, ΔK_{tip}

$$\Delta K_{tip} = \Delta K_{app} - \Delta K_b \tag{1}$$

where ΔK_{app} is the applied stress intensity factor range for a corner crack specimen geometry subjected to bending [*17, 18*], adjusted to account for temperature changes, and ΔK_b is the stress intensity factor range due to fiber bridging [*9*]. ΔK_{tip} is solved for iteratively at each cycle of the experiments. To allow direct comparison with fatigue crack growth data of the monolithic matrix material [*19*], a modulus correction is made to determine the effective stress intensity factor in the matrix, ΔK_m [*20*]

$$\Delta K_m = \sqrt{\frac{E_m}{(1-V_f)E_L}}\Delta K_{tip} \tag{2}$$

where E_m is the modulus of the matrix, E_L is the longitudinal composite modulus and V_f is the volume fraction of fibers.

Results and Discussion

In order to determine the effects of temperature, applied stress, tensile and compressive hold times, loading rates and cycle time, the matrix of tests shown in Table 1 was completed. The phasing of the applied temperature and stress was varied to study the effect of isothermal, and in-phase and out-of-phase TMF conditions on crack growth. Experiments were conducted on ten specimens: five were subjected to isothermal fatigue and five were subjected to TMF. This resulted in a total of twenty fatigue crack growth experiments: specimen–1 refers to Crack 1 on the top surface and specimen–2 refers to Crack 2 on the bottom surface.

Table 1 - *Test conditions applied to each specimen.*

Specimen	Test Type	Tmax (°C)	Tmin (°C)	ΔT (°C)	Tmean (°C)	σ_{max} (MPa)	σ_{min} (MPa)	Cycle Period (s)	Applied Cycles
133-2	Isothermal	371	371	0	371	462	-416	412	2048
		482	482	0	482	389	-478	412	508
		538	538	0	538	437	-403	412	480
133-3	Isothermal	427	427	0	427	484	-485	412	2046
133-4	Isothermal	538	538	0	538	433	-425	412	3549
133-5	TMF	538	204	334	371	474	-481	1672	536
133-6	TMF	538	371	167	454	498	-499	1032	1076
133-7	TMF	454	288	166	371	468	-476	1032	1507
133-9	TMF	454	288	166	371	445	-457	2652	602
133-10	TMF	538	371	167	454	383	-429	1032	1002
133-11	Isothermal	427	427	0	427	541	-556	1984	880
133-12	Isothermal	427	427	0	427	481	-489	2332	588

Fatigue Crack Growth Results

The average crack growth rates of isothermal 538°C tests 133-4-1, 133-4-2, 133-2-1 and 133-2-2 are compared with monolithic matrix crack growth rate curves in Figure 9. For a given stress intensity factor range, ΔK_{app}, applied to the composite, the resulting crack growth rate in the monolithic matrix would be about an order of magnitude higher than it is in the composite. Note for a given ΔK, the solid symbols fall below the elevated temperature monolithic matrix curves. Due to fiber bridging, the composite exhibits significant reduction in crack growth rates, resulting in rates close to the monolithic matrix tested at 23°C. The open symbols represent the calculated effective stress intensity factor range in the matrix, ΔK_m. As shown in Figure 9, plotting the average crack growth rates as a function of ΔK_m shifts the data to where it is close to the monolithic matrix data at 650°C.

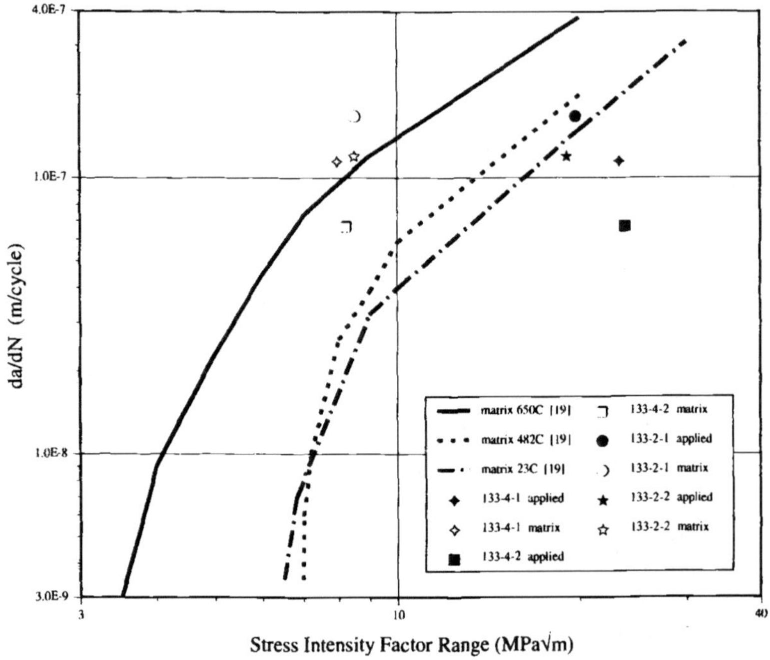

Figure 9 - *Comparison of isothermal 538°C average crack growth results with matrix fatigue crack growth curves from 23°C to 650°C.*

The average crack growth rates of all of the TMF tests are compared with the monolithic matrix crack growth rate curves in Figure 10. TMF tests that end with a -1 are in-phase experiments (open symbols) and those that end with a -2 are out-of-phase experiments (solid symbols). The average crack growth rate for each experiment is plotted both as a function of ΔK_{app} and ΔK_m in Figure 10(a) and Figure 10(b), respectively. In all of the tests, the composite showed a reduction in crack growth rate over the matrix material. Crack growth rates for ΔK_{app} were always lower than all of the matrix crack growth curves. Plotting the crack growth rates as a function of the ΔK_m shifts the data towards the matrix crack growth curves. By comparing in-phase and out-of-phase experiments on the same specimen, Figure 10 shows that out-of-phase TMF tests have a higher stress intensity factor range than their in-phase counterpart tests. This is due to the larger effective applied stress range [9]. Using the effective applied stress range for the TMF tests gives a correlation of the in-phase and out-of-phase data: a higher stress intensity factor range is calculated for the out-of-phase tests, which always have higher crack growth rates than the in-phase tests of same conditions. Comparing in-phase tests 133-6-1 and 133-10-1 using either ΔK_{app} or ΔK_m shows that a higher crack growth rate occurs in test 133-6-1, due to the higher applied stress range. This is also true when comparing 133-6-2 with 133-10-2, a 20% difference in applied stress range results in over an order of magnitude difference in the crack growth rates. Briefly, other observations that can be made are: 1) crack growth rates increase with temperature range and mean temperature, and 2) in isothermal fatigue compression hold times cause higher crack growth rates than tension hold times while the reverse is true in TMF.

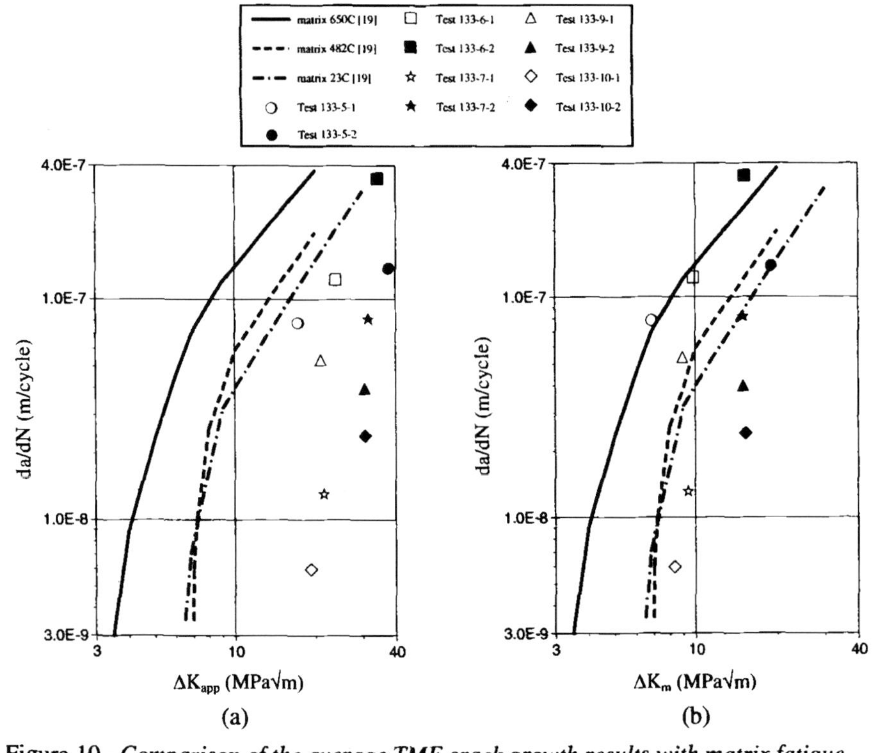

Figure 10 - *Comparison of the average TMF crack growth results with matrix fatigue crack growth curves for (a) applied stress intensity factor range, (b) effective stress intensity factor range in the matrix.*

Individual Cycle Results

In addition to collecting electrical potential data during the hold periods at maximum and minimum stress to determine crack dimension changes with cycles, electrical potential data was also collected periodically throughout entire individual loading cycles during each test to determine how the crack opens and closes during a single cycle. The individual cycle responses are a representation of the predicted crack dimension changes during loading cycles. Actual physical increases in the crack dimensions can only be accurately measured at maximum load. Changes in the predicted crack dimensions during individual loading cycles represent the opening and closing of the crack rather than a physical change in the crack dimension. The individual cycle responses were found to be uniquely related to the type of experiment, due to various mechanisms that are dependent on thermomechanical loading conditions. The responses in the length and depth directions are similar, but more sensitive in the depth direction due to loading in bending.

Test 133-2-1 was conducted isothermally at three progressively higher temperatures. The individual cycle response is shown in Figure 11. During all of the cycles applied at 371°C, the predicted crack depth did not vary with applied stress and the crack depth did not increase. After raising the temperature to 482°C, an immediate change in the individual cycle response was observed. Between zero stress and maximum stress, the predicted crack depth does not vary. Unloading into compression from zero stress results in a reduction in predicted crack depth, corresponding to crack closing. Holding for 30 s in compression results in an additional small reduction in predicted crack depth. Loading from minimum stress to zero stress results in an increase in predicted crack depth, corresponding to the crack opening back up. The loop slowly becomes smaller and closes over the 508 applied cycles. Once the temperature is increased again to 538°C, another immediate change in the individual cycle response is seen.

Examination of the individual cycle data from the different types of experiments performed shows traits that are common to all of the experiments of a given type. Therefore it is possible to show a typical response as a schematic of the change in predicted crack depth with applied stress during an individual loading cycle. Figure 12 shows schematic responses for isothermal tests, slow-fast and fast-slow isothermal tests, and in-phase and out-of-phase TMF tests. A detailed description of the mechanisms responsible for these responses is found in Ref. 9. The overall ranking of crack closure, or individual cycle loop size, corresponds with the fatigue crack growth results: out-of-phase TMF is the most severe, followed by isothermal fatigue and in-phase TMF.

Conclusions

A fully automated thermomechanical fatigue test system that is capable of applying prescribed bending loading cycles and monitoring crack growth throughout cycling as well as crack closure mechanisms during individual cycles has been developed. This system and the various experimental and analytical techniques associated with it have been described in this paper. The predicted crack dimensions compared well with the final measured crack dimensions and the predicted crack dimension changes during individual cycles has led to a better understanding of the crack growth processes occurring in titanium matrix composites during various thermomechanical bending loading conditions.

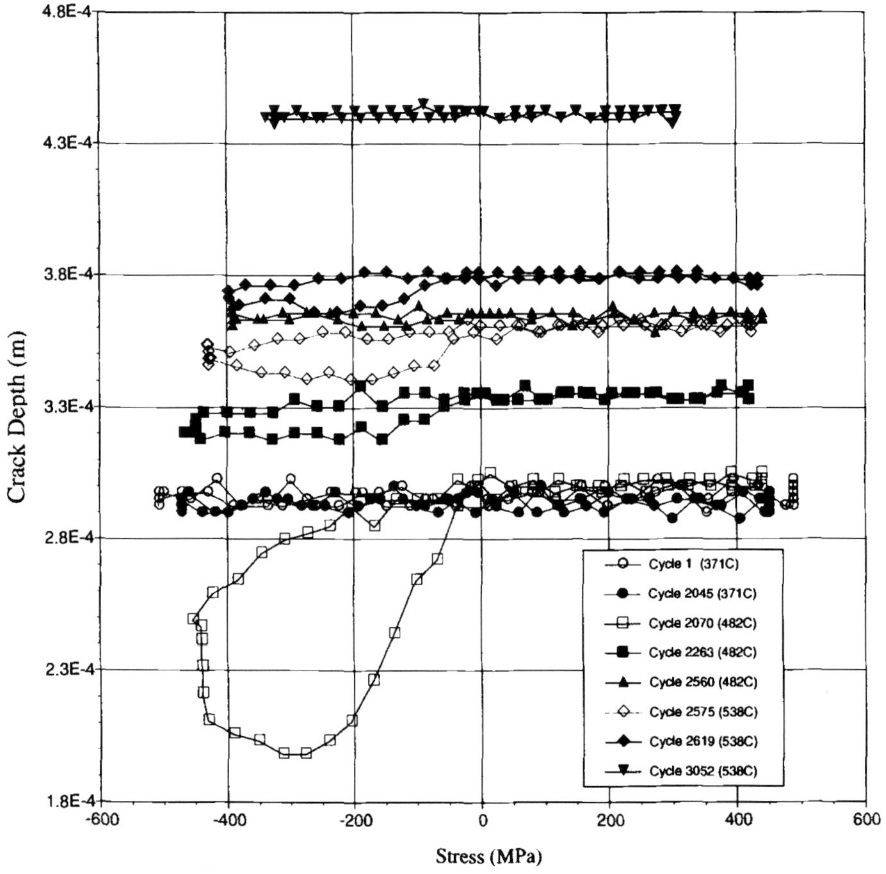

Figure 11 - *Individual cycle response of test 133-2-1, isothermal fatigue: 2048 cycles at 371 °C, 508 cycles at 482 °C, 480 cycles at 538 °C, 30 s hold in tension, 30 s hold in compression.*

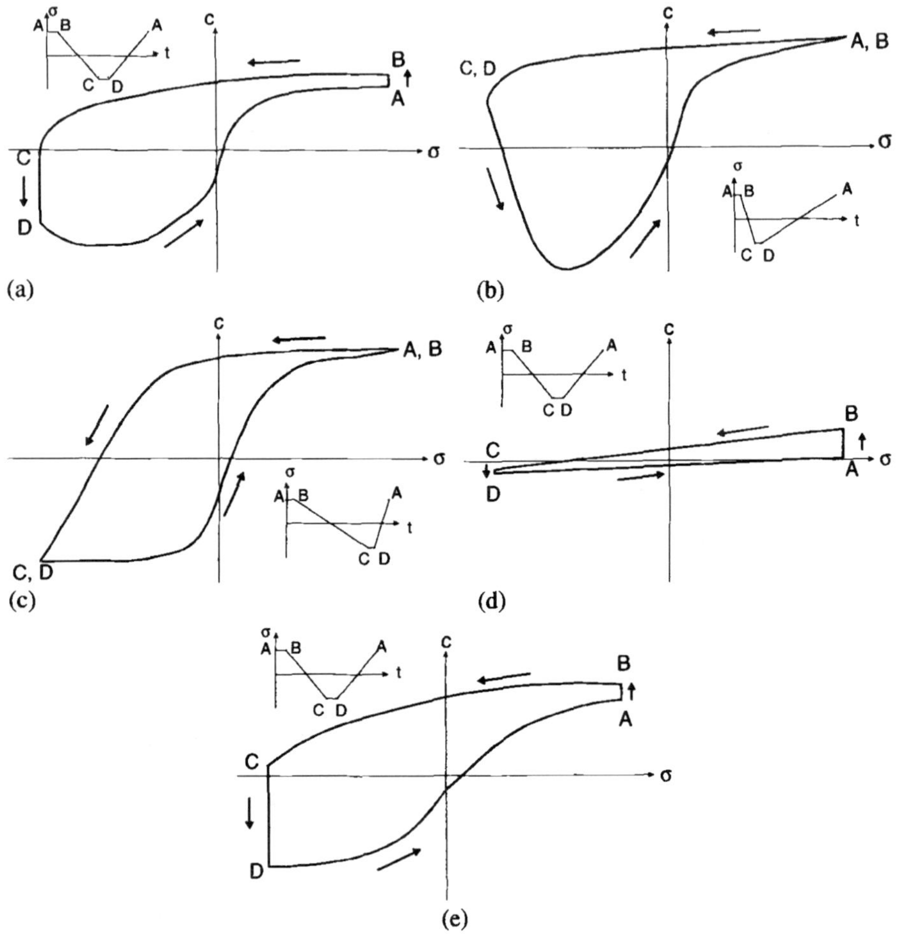

Figure 12 - *Schematic of the change in predicted crack depth as a function of stress during an individual loading cycle: (a) isothermal, (b) slow-fast isothermal, (c) fast-slow isothermal, (d) in-phase, and (e) out-of-phase.*

Acknowledgments

This work was performed at the Mechanics of Materials Laboratory of Rensselaer Polytechnic Institute. Support provided by General Electric Corporate Research and Development Center is gratefully acknowledged. We also wish to thank M. F. Henry and G. M. Roe (retired), both of General Electric Corporate Research and Development Center, for helpful discussions and technical guidance. The titanium matrix composite material was kindly provided by T. Nicholas of the Air Force Research Laboratory.

References

[1] Mall, S. and Portner, B., "Characterization of Fatigue Behavior in Cross-Ply Laminate SCS6/Ti-15-3 Metal Matrix Composites at Elevated Temperature," *ASME Journal of Engineering Materials and Technology*, Vol. 114, 1992, pp. 409-415.

[2] Majumdar, B.S. and Newaz, G.M., "Thermomechanical Fatigue of an Angle-Ply Metal Matrix Composite," *Composite Materials: Fatigue and Fracture, (Third Volume), ASTM STP 1110*, T.K. O'Brien, Ed., American Society for Testing and Materials, Philadelphia, 1991, pp. 732-752.

[3] Russ, S.M., Nicholas, T., Bates, M., and Mall, S., "Thermo-mechanical Fatigue of SCS-6/Ti-24Al-11Nb Metal Matrix Composites," *Failure Mechanisms in High Temperature Composite Materials*, ASME AD-Vol. 22 and AMD Vol. 122, 1991, pp. 37-43.

[4] Mall, S. and Schubbe, J.J., "Thermo-Mechanical Fatigue Behavior of a Cross-Ply SCS-6/Ti-15-3 Metal Matrix Composite," *Composites Science and Technology*, Vol. 50, 1994, pp. 49-57.

[5] Mall, S., Hanson, D.G., Nicholas, T., and Russ, S.M., "Thermomechanical Fatigue Behavior of a Cross-Ply SCS-6/β21-S Metal Matrix Composite," MD-Vol. 40, *Constitutive Behavior of High-Temperature Composites*, ASME, 1992, pp. 91-106.

[6] Ball, D.L., "An Experimental and Analytical Investigation of Titanium Matrix Composite Thermomechanical Fatigue," *Thermomechanical Fatigue Behavior of Materials: Second Volume, ASTM STP 1263*, Michael J. Verrilli and Michael G. Castelli, Eds., American Society for Testing and Materials, West Conshohocken, PA, 1996, pp. 299-330.

[7] Jeng, S.M. and Yang, J.-M., "Damage Mechanisms of SCS-6/Ti-6Al-4V Composites Under Thermal-Mechanical Fatigue," *Materials Science and Engineering*, Vol. A156, 1992, pp. 117-124.

[8] Neu, R.W. and Nicholas, T., "Effect of Laminate Orientation on the Thermomechanical Fatigue Behavior of a Titanium Matrix Composite," *Journal of Composites Technology & Research*, JCTRER, Vol. 16, No. 3, July 1994, pp. 214-224.

[9] Antoun, B.R., *Crack Propagation From Initial Defects in Metal Matrix Composites During Thermomechanical Cycling*, Ph.D. Thesis, Rensselaer Polytechnic Institute, Troy, NY, May 1998.

[10] Murray, M.L., *Experimental Methods and Damage Observations in Bending Fatigue Testing of Metal Matrix Composites*, Masters Thesis, Rensselaer Polytechnic Institute, Troy, NY, May 1991.

[11] Catlin, W.R., Lord, D.C., Prater, T.A., and Coffin, L.F., "The Reversing dc Electrical Potential Method," *Automated Test Methods for Fracture and Fatigue Crack Growth, ASTM STP 887*, American Society for Testing and Materials, Philadelphia, 1985, pp. 67-85.

[12] Prater, T.A. and Coffin, L.F., "Part-Through and Compact Tension Corrosion Fatigue Crack Growth Behavior of Carbon Steel in High-Temperature Water," General Electric CRD Report No. 81CRD159, July 1981.

[13] "Environmental Crack Growth Measurement Technique," Electric Power Research Institute Report, EPRI NP-2461, Project 2006-3, prepared by General Electric Co., Corporate Research & Development, Schenectady, NY, November 1982.

[14] Prater, T.A., Catlin, W.R. and Coffin, L.F., "Environmental Crack Growth Measurement Techniques," EPRI Project RP 2006-3, Interim Report, October 1983.

[15] Prater, T.A., Catlin, W.R. and Coffin, L.F., "Environmental Crack Growth Measurement Techniques," General Electric CRD Report No. 85CRD167, September 1985.

[16] Roe, G.M., Private Communications, January 1992.

[17] Newman, J.C. Jr. and Raju, I.S., "Stress-Intensity Factor Equations for Cracks in Three-Dimensional Finite Bodies Subjected to Tension and Bending Loads," *NASA TM-85793*, April 1984.

[18] Raju, I.S. and Newman, J.C. Jr., "Finite Element Analysis of Corner Cracks in Rectangular Bars," *NASA TM-89070*, June 1987.

[19] Ghonem, H., Wen, Y. and Zheng, D., "Effects of Temperature and Frequency on Fatigue Crack Growth in Ti-β21S Monolithic Laminate," *Materials Science and Engineering*, Vol. A161, 1993, pp. 45-53.

[20] McCartney, L.N., "Mechanics of Matrix Cracking in Brittle-Matrix Fibre-Reinforced Composites," *Proceedings, Royal Society of London*, A 409, 1987, pp. 329-350.

Full-Scale Testing

Roy L. Hewitt[1] and Arlin Nelson[2]

Data Trend Monitoring and End Level Verification – Tools to Reduce Data Storage in Full-Scale Aircraft Fatigue Tests

Reference: Hewitt, R. L. and Nelson, A., "**Data Trend Monitoring and End Level Verification – Tools to Reduce Data Storage in Full-Scale Aircraft Fatigue Tests,**" *Applications of Automation Technology in Fatigue and Fracture Testing and Analysis: Fourth Volume, ASTM STP 1411,* A. A. Braun, P. C. McKeighan, A. M. Nicholson, and R. D. Lohr, Eds., American Society for Testing and Materials, West Conshohocken, PA, 2002.

Abstract: The reduction in cost of large data acquisition systems and the ease of taking data has encouraged analysts to demand more and more data from full-scale structural tests. However, this can lead to excessive data storage requirements. Furthermore, the amount of data can sometimes result in data handling difficulties, which prevent convenient and timely examination of the data.

New trend monitoring and load checking systems are described that can reduce the amount of data that must be stored. These systems also provide the user with very rapid feedback during a test of any changes in either the structure or the testing system. Examples of the use of these features in a full-scale fatigue test of a fighter aircraft wing are provided.

Keywords: aircraft, fatigue, full-scale testing, data acquisition, monitoring

The reduction in cost of large data acquisition systems and the ease of taking data has encouraged analysts to demand more and more data from full-scale structural tests. It is not uncommon to require more than 1000 strain gages on a fatigue test article. However, if data are required at every end point, this can cause problems with respect to data storage because of the immense amount of data that continuous end point data collection entails (typically over 100 GB). While this will become less important as costs of data storage decrease, the subsequent data handling will still cause significant problems. In addition, there are sometimes requests to pause at each end point to ensure the data are all taken at the same load level, and this impacts testing time. The problems are exacerbated by the current trend toward very long load histories [*1*].

This raises the question of what the real purpose of the data is. Initially, data must be taken at every independent load condition so that a strain spectrum can be derived for various critical locations on the structure. Additionally, if there is hysteresis in the

[1] Senior research officer, Institute for Aerospace Research, National Research Council of Canada, Montreal Road, Ottawa, Ontario, Canada, K1A 0R6.

[2] Systems engineer, Aerospace Systems and Consulting, MTS Systems Corporation, 14000 Technology Drive, Eden Prairie, MN 55344.

structure, it may be necessary to take data at every end point in the first loading block to account for variations in strain caused by the previous end point. Thereafter, strain data are only important if they change because this signifies either a problem with the gage, the loading or the structure. Conventionally, this is done by analyzing end point data off-line, but because this is very difficult with such large amounts of data, it is often only done after a problem is encountered.

Hewitt and Rutledge [*2*] described a simple trend monitoring system that was implemented on a full-scale fatigue test of a Tutor aircraft that reduced the amount of data that had to be stored. Using digital input and output lines for communication between the control and data acquisition systems allowed data on specific segments of the loading history to be collected and checked against predefined limits throughout the test. However, data were only checked at one particular load condition and each time an error limit was exceeded it was necessary to examine past trends from plots generated off-line on another system.

This paper describes a new trend monitoring system that allows limit checking on a very large number of different end points with individual limits for every data channel assigned to each end point. Thus data can be checked on any load condition that is critical for any section of the structure. The system also automatically plots recent trend data for any limit exceedence.

The other type of data that is often requested is load data at every end point so that the analyst or certifying authority can check that every load condition has been applied correctly. Again, the problem is the extensive storage facilities required and the difficulty of checking the data. In addition, checking the load levels at some nominal end point does not ensure that the loads were not exceeded just before or just after the recording point. It is therefore more appropriate to develop on line end level verification and only record selected end points in conjunction with the data checking.

A new software system that has been developed to check the loads throughout the loading cycle is described. This system will notify the operator whenever an end point is missed and keep track of all missed end points. Examples of the use of these features in a full-scale fatigue test of a fighter aircraft wing are also included.

Trend Monitoring System

Requirement

The system described by Hewitt and Rutledge [*2*] only allowed checking on one load condition. In a full-scale aircraft fatigue test, there are many different critical locations and these are generally sensitive to different loading actions. For example, on a wing test, there will be some locations that are sensitive to wing root bending and others that are sensitive to control surface hinge moment. Since these two loads are not generally dependent, taking data on a load line that produces a significant strain at the location sensitive to wing root bending may result in a very small strain at the location sensitive to control surface hinge moment. Then any changes at the second location will be difficult to observe. Thus the system must be able to check strains on at least as many load conditions as there are independent loading actions. For a wing, this might be of the order of thirty, while for a full aircraft it may be over one hundred.

With the large number of data channels on modern tests, the user could not manually check the data trends, even if there were only one loading condition. The system must therefore be able to check the measured data against some defined limits and alert the operator when the limits are exceeded. Clearly there must be independent limits for each gage for every load condition. Again, because there will then be a very large number of limits to be input to the system, some method of bringing in limits from a spreadsheet application or of generating the limits automatically is required.

The trends in the data can be observed by plotting as some function of time. While the index of the data scan could be used, a more logical parameter in a fatigue test would be flight hours or number of flights. The software should therefore have this functionality.

It has been shown [*2*] that strain gages should not be re-zeroed on a regular basis during a fatigue test because the strain gage output at the nominal zero load condition contains valuable information. An example was given of a gage that showed considerable scatter at zero load but almost none at the required load condition. This was caused by hysteresis in the structure. If the gage had been re-zeroed each day, this hysteresis effect would not have been observed and the scatter would have shown up incorrectly in the at-load data. However, strain gages can deteriorate during a test, particularly if they are used in high strain locations, and zeros can tend to drift. It is therefore important for the system to be able to plot both the at-load data and the zero-load data on one plot, so that changes due to zero drift can be isolated.

Software Implementation

The trend monitoring software was designed for use with the current MTS Aero-90™ hardware and software and implemented on a high-end business PC running the Windows NT operating system. A PC was chosen for cost, ease of use, and availability while Windows NT was chosen as the operating system because it is the MTS corporate standard. The software was designed using a Microsoft Visual Basic user interface that manages several COM (component object model) components written in C++. The COM components manage all data manipulation and perform the majority of the tasks that occur behind the user interface. Olectra Chart, from KLGroup, was used for the plotting software for the graphs. This is a lower level piece of software that is managed by C++ objects to display data.

The PC that is running the trend monitoring system communicates with the Aero-90™ system via an Ethernet connection and is designed to be an integral part of the load controller and data acquisition system. When the software is launched it looks for an Aero-90™ data acquisition test to connect to. Once the data acquisition test is connected the trend monitoring software reads information about the test channels automatically. The information read from the Ethernet link informs the trend monitoring software how many load control channels and data acquisition channels are in the test. The channel description, engineering units, and channel full-scale values are also read for each channel.

The key features of the software are designed to make the setup and use of the software as easy as possible. This is further enhanced by some global settings that can be adjusted to tailor the way the software works and presents data. These global settings adjust default parameters that control how plots are displayed and how limits are calculated.

Software Interface

The Trend Limits table shown in Figure 1 is the first window that the user interacts with to set up the trend monitoring software. This window contains all channels (both control channels and data acquisition channels) in the test by default. The user can define limits for any of the load conditions in the test.

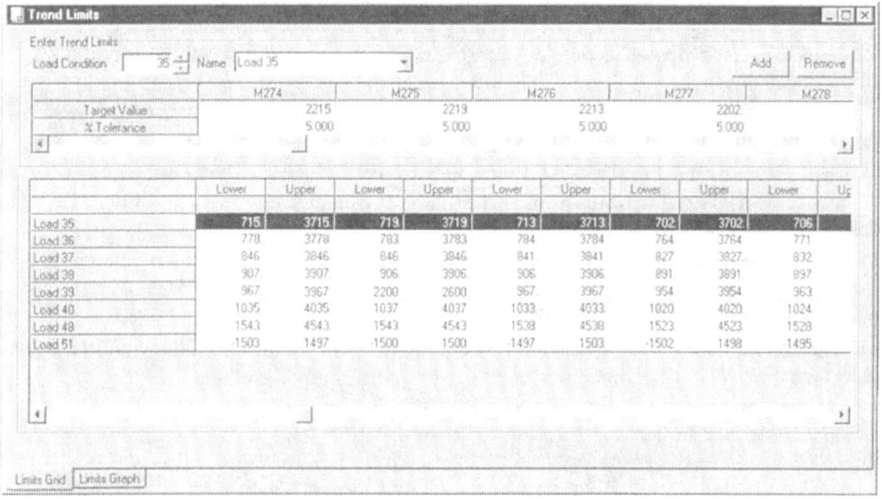

Figure 1 – *Trend limits table window.*

All data that are collected in the data acquisition test are sent to the trend monitoring computer. This computer stores the data for the load conditions that are listed in the trend limits table. If at any time during the test a load condition is added to the trend limits table, then the data that are collected at that load condition are stored from that time on.

The software allows the user to cut and paste trend limits into the limit definition table from various spreadsheet applications. If the task of predicting the value and limits for each test channel at each load condition requires too much time, then the software can automatically generate the limits at the time the data for a load condition are first stored. This option requires the user to input a tolerance range for all channels that are to have the limits automatically generated. After an input tolerance has been specified for channels at a specific load condition, then the next time that data are taken at that load condition, the software will automatically generate the limits for all channels that have a specified tolerance. The tolerance can be automatically generated from either percent of full scale or percent of the value read. If the data are around zero the later method may generate limits that are very close together. If a load condition is input by the user and no limits are applied (either manually or automatically generated) then the data are stored for that load condition, but obviously there will be no need for comparison to other readings for this is the same as having infinite limits. However, these data are available for plotting and may be of value if the limits are defined in the future for this load condition.

Once the limits are set, the software can display the limits in a bar graph for all of the channels as shown in Figure 2.

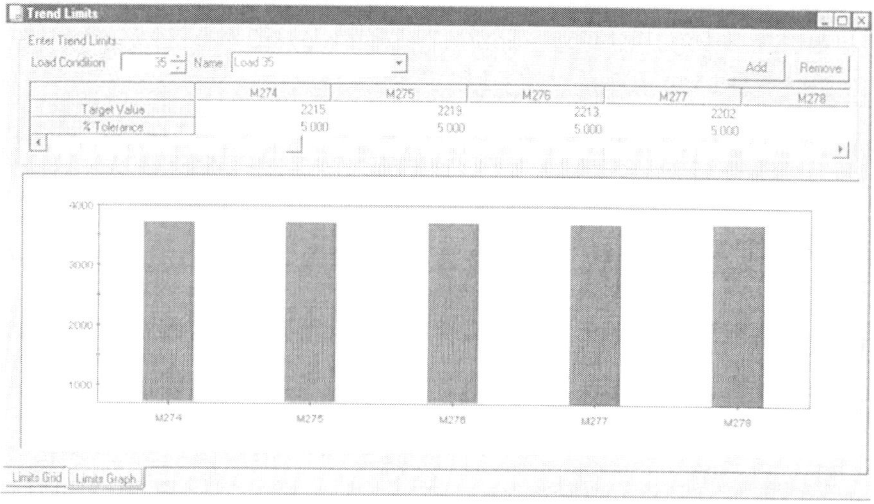

Figure 2 – *Bar graph window.*

If a limit is exceeded, the trend monitoring software will generate a message that is displayed in the log as shown in Figure 3. The information displayed shows both the description of the limit exceeded and its value. The log window also displays the status of the Ethernet connection. The button at the lower right of the window by the "On Line " label indicates that the trend monitoring software is communicating with the data acquisition test if it is green. If the connection fails, the button will turn red so that the operator knows that the connection has failed.

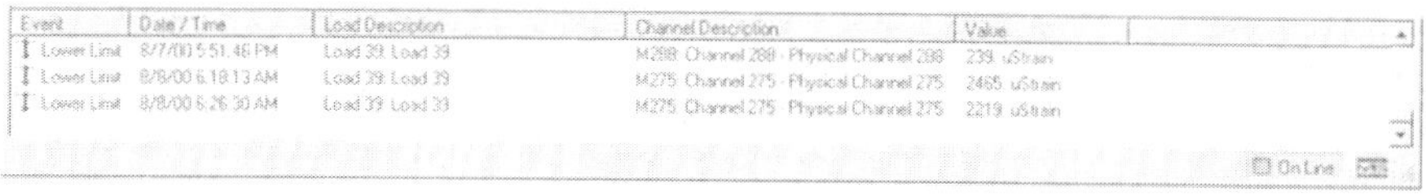

Figure 3 – *Log file window.*

The log entries also have an icon at the left of the information describing what caused the limit to be exceeded. Double clicking on the icon in the log window launches a Quick Plot, shown in Figure 4, that will display previously stored data for the channel and load condition with the error for a user selectable amount of data samples prior to the error. The plots can be manipulated using standard windows operations to change headers, axis values, legends etc. It can be seen in Figure 4 that the data for the 0g condition are slowly increasing while the data for load condition UID08179 is very erratic. This suggests a problem with the strain gage and this was determined to be the case by comparing data for other load conditions.

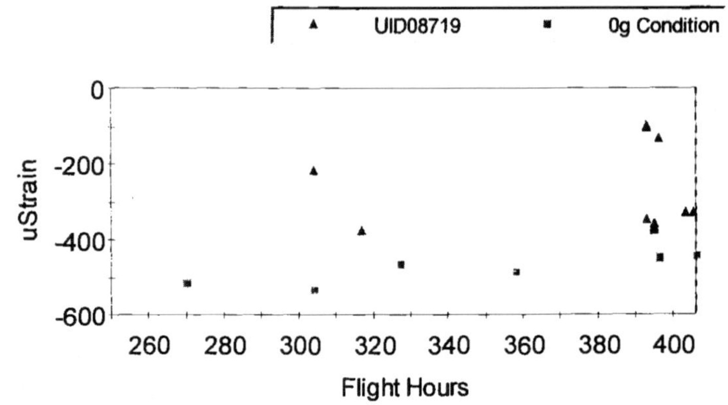

Figure 4 – *Quick plot window.*

Figure 5 shows a typical normal data plot of a strain output channel with zero-load (0g condition) and at-load data (load condition UID08719) plotted against flight hours. Both the zero load and at-load data show some variation. The variation in zero load data was observed by Hewitt and Rutledge [*2*] and attributed to hysteresis in the structure. However, the at-load data is not quite as consistent as previously observed. This is a result of the binning process. The loads used in [*2*] represented complete manoeuvres and the load condition on which data were taken was always preceded by the same load conditions. However, with the binning process, the occurrence of a particular load condition (UID08179 in this case) is not always preceded by the same load condition.

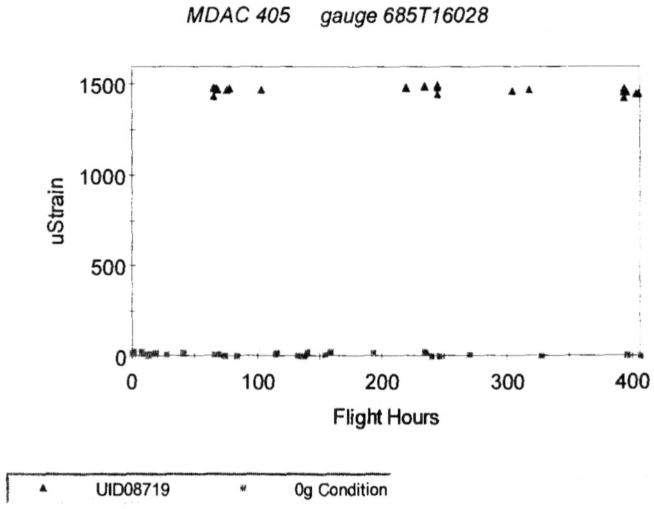

Figure 5 – *Normal plot window.*

The hysteresis of the structure then results in slightly different strains depending on what the previous load condition was. However, the load sequence is repeated every 326 flight hours, so the data near 400 flight hours corresponds to the same sequence of loads as the data near 74 flight hours. Comparing the data for these two groups shows a consistent pattern, confirming that the hysteresis is responsible for the variations in the at-load data.

Other key features in the trend monitoring system allow it to be integrated into the test by communicating directly with the load controller and the data acquisition system. Then, when a trend limit is exceeded, the load controller can react in a programmed way. The communication link to the load controller and data acquisition system also provides the opportunity for a handshake to ensure that data that are intended to be collected by the trend monitoring system have been collected. This ensures that no data are missed and that the data checking cannot be turned off inadvertently.

Selection of Load Lines for Trend Monitoring

The National Research Council of Canada is currently performing a full-scale fatigue test of an F/A-18 wing for the Canadian and Australian Air Forces. A brief description of the test set up and spectrum development is provided in Reference 1 and an overall view of the wing is shown in Figure 6.

Figure 6 – *F/A-18 wing test.*

Because the test includes the effects of trailing edge buffet, there are over 160 000 load lines in the full spectrum. Since the loads were developed directly from recorded flight data, these load lines are all unique. However, many of the load lines are quite similar and it was decided to group similar load lines and use one representative load line for each group for application to the test [*3*]. This was done in order to reduce the computation time for the calculation of actuator loads and avoid breaking the spectrum into several blocks as the hardware limited the number of independent load lines to 50 000. This grouping, or binning, of load lines resulted in a reduction in the number of independent load lines to about 49 000.

Binning of similar load conditions enhances the use of trend monitoring. If load conditions only occur once per block, there is a long time between checking of each condition. However, if a monitored load condition occurs 10 times in every block, the data is checked 10 times more often. This allows the test engineer to react more rapidly to any changes in the structure that are evident from changes in strain (or other monitored parameter).

For the current wing test, 35 independent loading actions were identified as influencing various parts of the wing structure. Of these, there were 23 primary loads that were easily determined from the spectrum loads. These included bending moments, shears and torques at the wing root, wing fold and wing tip, the control surface hinge moments and root shears and some stores loads. Load lines for trend monitoring were then selected for each of these loading actions. This was done by finding the bins that contained the highest and lowest load cases for each of the 23 primary loads with a bin population of more than 9. This resulted in 31 unique load cases.

Since the extreme load cases may be more sensitive to changes in the structure, these load cases were also selected for monitoring, even though they occur less frequently. This resulted in an additional 35 load cases for a total of 66 unique load cases for trend monitoring.

Load Checking System

Requirement

When interpreting the results of a fatigue test, it is very important to know exactly what loads were applied. One method to determine this might be to measure every load at every end point. However, besides requiring large amounts of data storage, this method has significant problems, mainly related to the time at which the loads are measured.

In a full-scale fatigue test there are a large number of actuators applying load. While all the command loads are synchronized, the applied loads will not be exactly in phase as some error is required in order to drive the actuators. Most actuators will tend to lag the command, but there will be some actuators that lead the command. If the loads are measured at the command end point, some loads will not have reached their peak yet while others may be beyond the peak. The only way to measure the peak loads on each actuator would be to have a peak detection system on every feedback channel. This would require excessive computing power and would still be of little use in analysis because the loads would all be measured at different times. It would be more useful to the analyst to be able to guarantee that the command loads were achieved with some defined tolerance.

Modern control systems have a number of error detectors that can assist in providing this guarantee. Typically, a system will be set up to warn the operator if the feedback error exceeds some inner error band and shut down the test if it exceeds an outer error band. However, the end point accuracy is much more important than the accuracy between end points and the error limit must be set for this transition error. This inner error detector is usually set at about 5%, and so the operator can only guarantee that the loads have been achieved within 5% of the command end levels.

Modern control systems also use some kind of null pacing software that can hold a test at an end point until all control channels are within a specified limit, typically 1 or 2%. Then the operator can guarantee that the loads at the end points were with this null pacing limit. However, although pacing software will not proceed until the load on every channel is within limits, it does not prevent or record any overshoot. Thus the operator only knows that the loads were within tolerance at the end points but may have overshot the peak by as much as just under the inner error limit. Since a 5% over-peak can have a significant effect on fatigue life, this is not an adequate guarantee.

The requirement is therefore for a system that will track all the applied loads and report any occurrences where the applied loads exceed the command end points (targets) by more than some specified error.

Software Implementation

Before describing the implementation of the target validation system, it is necessary to define the term "End of Segment (EOS)". The end of segment is when the command has reached the endpoint and the feedback level for every channel is within the null pace limits at the same time.

The system has been designed to monitor every channel on every transition from the end of the last segment to the end of the current segment. Any event such as an over-peak that exceeds the null pacing limit and would introduce additional damage is recorded and associated with that segment. This is achieved by defining a box around each segment that exceeds the peak by the null pacing limit and is lower than the valley by the same amount. This target validation window is illustrated in Figure 7 for three consecutive segments.

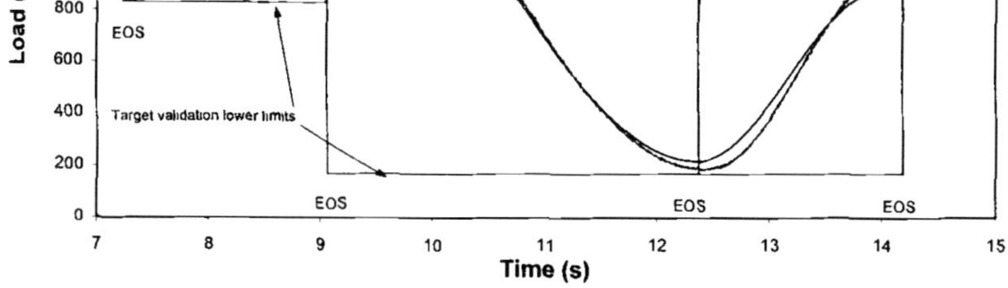

Figure 7 – *Target validation windows.*

During the transition from EOS to EOS there is a possibility of overloading and underloading in each segment. For example, at the end of the first segment in Figure 7, the feedback is higher than the command peak and an over-peak is recorded and associated with this segment. At the end of the second segment, the feedback is lower than the command valley: this is defined as an under-peak, although in this instance it is not recorded as it does not exceed the target validation lower limit for the segment.

It is possible for the same over-peak or under-peak to be recorded in two segments if a null pacing error occurs. For example, if the over-peak at the end of the first segment in Figure 7 did not fall within the target validation window at the end of the segment, a null pacing error would be recorded. The target validation window would then be exceeded as soon as the second segment began and the over-peak would be recorded for the second segment as well.

Software Interface

The Target Validation information is displayed in a typical runtime display bar chart as shown in Figure 8 for a four channel test. There are two different target validation counters that can be displayed. They are over-peak and under-peak counters and total over-peak/under-peak counters. The over-peak/under-peak counters can be reset by the operator. This is very useful for monitoring the system tuning process, as the number of exceedances in a given time period should decrease as the test tuning is improved. However, the total over-peak/under-peak counters cannot be reset. Thus there is a traceable record of all the over-peaks and under-peaks that occur during a test.

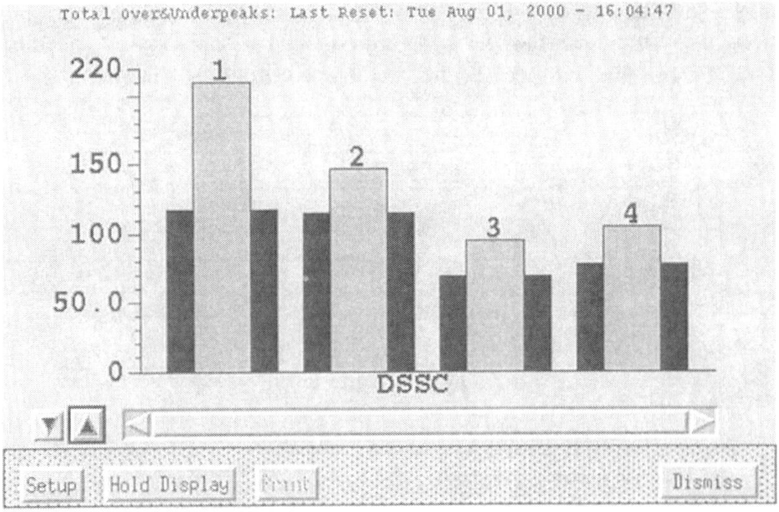

Figure 8 – *Total over-peak/under-peak window.*

The bar chart is updated once a minute. By clicking on the bar for one of the channels on the chart, it is possible to display the values for that channel as shown in Figure 9.

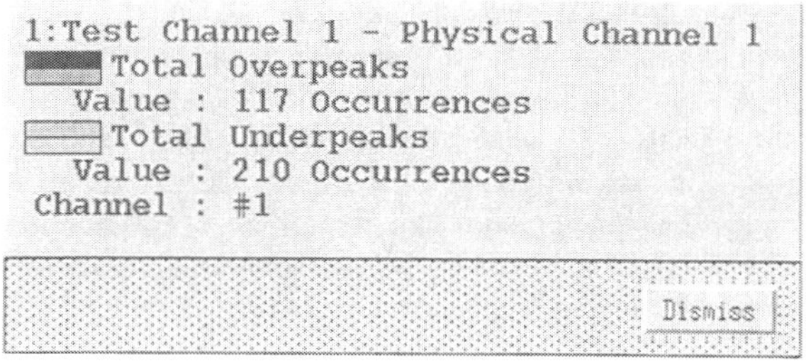

Figure 9 – *Specific channel details.*

When an over-peak or under-peak occurs in a segment, an event is logged in the test log at the end of the segment. This information includes the channel or channels, the over-peak and/or under-peak, the time, and the test counters.

Use of Load Checking System

The use of the load checking system can be best illustrated by reference to the current full-scale test of the F/A-18 wing referred to above. This test uses 63 servo-hydraulic actuators to apply the loads and the transition times for every segment (the time to go from one load condition to the next) were set based on the response of the slowest actuator for that transition [*2*]. Inner error (the difference between command and feedback) detectors were initially set at 2% of full scale for each actuator, static null pace error (the error at the EOS) limits are set at 1% and dynamic null pace error (the error at other than EOS) limits are set at 1.5%. If the inner error limit on any channel is exceeded, the test unloads to a zero g condition. If the static null pace error limit for any channel is exceeded, the system holds at that level until the error on that channel is below the static null pace limit. If this does not occur within 20 seconds, the test again unloads to a zero g condition. If the dynamic null pace error limit is exceeded for any channel, the system slows every channel on the test.

The test was started with the master time set at 500% so that the test was running 5 times slower than for the calculated segment times. This was to ensure that there would be no error limits exceeded with the initial tuning parameters. The proportional gains were gradually increased to the maximum that could be achieved without showing evidence of instability. The master time was then gradually reduced so that the test ran faster.

As test speed was increased, dynamic and static null pacing errors began to occur, together with over-peaks and under-peaks, on a number of channels for some segments. Dynamic null pacing errors were associated with insufficient fluid flow generally caused by insufficient valve opening. Because the proportional gain was already near the maximum, the dynamic proportional gain, which increases the proportional gain when the system is between end points where it is more stable, was increased. These adjustments were made as the errors occurred and the tuning soon reached a point where dynamic errors were not significant.

Static null pacing errors and over-peaks and under-peaks are slightly more difficult to deal with. They are generally caused by interactions between actuators and disturbances at end points caused by small amounts of backlash in the loading train or specimen. Once the proportional gains are as large as possible, the only possibility for improving the system response is by adding integral control. However, this can lead to oscillations and improving the response of one actuator can lead to a degradation in response of another. In addition, the system response changes throughout the loading block. It was therefore important to be able to monitor the response over period rather than on a per segment basis. The end level bar chart provided a simple way of doing this, both in terms of which actuators were causing problems and in terms of the number of occurrences as the test progressed.

At the current time, the test speed is limited by a group of three actuators near the wing tip that experience over-peaks and under-peaks if the test is run faster. These over-peaks are associated with disturbances at the end points as shown in Figure 10. The full scale for this channel is 5500 lbf so the error of about 50 lbf is less than 1% of full scale. It is believed that some minor changes in set-up will eliminate the backlash that is causing these errors. However, in the first block of 154 000 segments, less than 200 over-peaks or under-peaks occurred. This means that more than 153 800 load cases were applied within an accuracy of 1% of full scale on every actuator. For the other 200 load cases, most of the actuator loads were accurate to within 1% of full scale while one actuator was only accurate to within 2% of the full scale of that actuator. There were significantly more over-peaks and under-peaks than static null pacing errors. Thus for test systems without end level checking, it is possible to experience a large number of over-peaks or under-peaks without knowing.

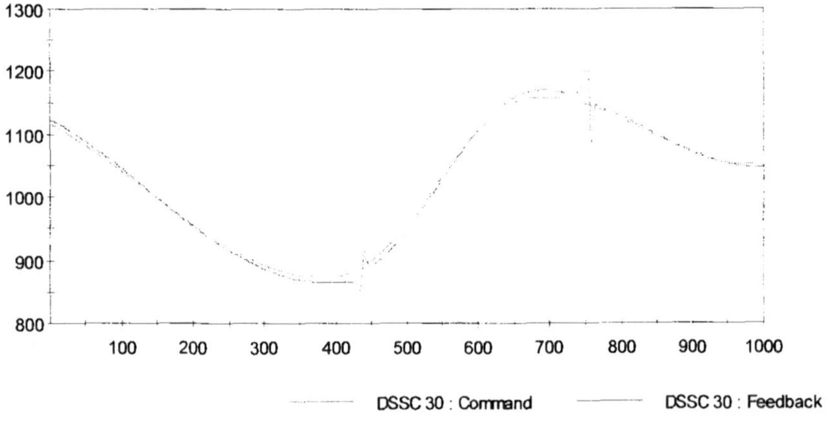

Figure 10 – *Command and feedback for channel 30.*

As these problems are resolved, the test speed will be increased. The aim is to run as fast as possible without any errors. By resetting the bar graph each day, it is simple to monitor any changes in the test response. If errors occur that cannot be eliminated, the test will be slowed down because it is very difficult to account for inaccurate loads on a multi-channel full-scale fatigue test. In a single channel test, it is possible to estimate the change in fatigue damage caused by any loading inaccuracies [*4*] because the stress in the specimen is a direct function of the load. For a similar analysis in a multi-channel test, it would be necessary to derive a relationship between the stress at every critical point in the structure in terms of all the actuator loads. Since one of the aims of the fatigue test is to define these critical points, this is not really feasible.

Conclusions

A trend monitoring system has been described that significantly reduces the amount of data that must be stored during a structural fatigue test and makes the data that are stored more useful. By continually monitoring the data on specific end levels and comparing with individual limits for every channel for each load condition, the system can notify the operator whenever there is a change in the structure, test rig or loading system that causes a change in any of the data channels.

A load checking system has been described that allows the documentation of exceedances of any channel peak or valley beyond a defined limit (typically 1 or 2%). This is considered to be far more useful than recording the feedback on every channel at some nominal end point and reduces the amount of stored data by several orders of magnitude.

Acknowledgments

The contributions of Dave Froslie and Terry Nelson of MTS Systems Corporation in developing the end level verification methodology are gratefully acknowledged. Thanks are also due to Mike Buth, Jeff Estum and Mike Sullivan of MTS Systems Corporation for developing the software.

References

[1] Hewitt, R. L., "Full-Scale Aircraft Fatigue Tests – Sandbags to Simulation," *Proceedings of International Conference on Mechanical Testing of Materials and Structures (English Edition),* S. Lee, J. W. Tang and D. S. M. Chor, Eds., The Hong Kong University of Science and Technology, 1998, pp. 1-42.

[2] Hewitt, R. L. and Rutledge, R.S., "Computer Applications in Full-Scale Aircraft Fatigue Tests," *Applications of Automation Technology to Fatigue and Fracture: Testing and Analysis: Second Volume, ASTM STP 1231,* C. Amzallag, Ed., American Society for Testing and Materials, Philadelphia, 1994, pp. 51-69.

[3] Nor, P. and Hewitt, R.L. "Development of Truncation and Binning Parameters for a Fighter Aircraft Wing Fatigue Test", to be published.

[4] McKeighan, P. C., Fess II, F. E., Petit, M. and Campbell, F. S, "Quantifying the Magnitude and Effect of Loading Errors during Fatigue Crack Growth Testing Under Constant Amplitude and Variable Amplitude Loading," *Applications of Automation Technology in Fatigue and Fracture Testing and Analysis: Fourth Volume, ASTM STP 1411,* A. A. Braun, P. C. McKeighan, A. M. Nicholson, and R. D. Lohr, Eds., American Society for Testing and Materials, West Conshohocken, PA, 2002.

Keith B. Smith,[1] Edward S. Parker,[2] and Darrell J. Iler[3]

Railcar Service Spectra Generation for Full-Scale Accelerated Fatigue Testing

Reference: Smith, K. B., Parker, E. S., and Iler, D. J., **"Railcar Service Spectra Generation for Full-Scale Accelerated Fatigue Testing,"** *Applications of Automation Technology in Fatigue and Fracture Testing and Analysis: Fourth Volume, ASTM STP 1411,* A. A. Braun, P. C. McKeighan, A. M. Nicolson, and R. D. Lohr, Eds., American Society for Testing and Materials, West Conshohocken, PA, 2002.

Abstract: North American freight railroads collectively use a computer-controlled servohydraulic test bed called the Simuloader to perform full-scale multiaxial fatigue tests of railcars at accelerated rates. It is used to verify analyses, reveal design deficiencies, test modifications and provide safety from failures not accounted for in standard specification tests. While material coupon testing can address very precisely and directly the science of mechanical fatigue, full-scale testing is often used to address gross design, manufacturing and production-related fatigue issues. The input waveforms used to animate this machine are typically created from combinations of both time histories and cycle counted data. Simulations are accelerated through the utilization of only those events that produce significant stresses on the structure. This paper describes the test bed and its vehicle interfaces, the stress analysis of a test car, acquisition of service input and car body response data, drive file development techniques and the iterative process of tuning the machine/railcar system to improve global response accuracy. The complex data gathering and reduction process associated with creating and tuning input drive files for this railcar test bed is illustrated using data from experiments recently performed by the railroad industry.

Keywords: accelerated test, car body fatigue, freight train, instrumentation, multiaxial loads, railcar, rainflow histogram, ride quality, strain measurement, servohydraulic test bed, simulation, Simuloader, stress analysis, spectrum, transducer, truncation

[1]Senior Mechanical Engineer, Ingersoll-Rand Company, 3899 Hancock Expressway, Security, CO 80911.
[2]Senior Mechanical Engineer, Colorado Railcar Manufacturing, LLC, 1011 14th Street, Fort Lupton, CO 80621.
[3]Senior Mechanical Engineer, Transportation Technology Center, Inc., 55500 DOT Road, Pueblo, CO 80110.

Railcar Test Bed and Experimental Background

The Simuloader: A Full-Scale Railcar Test Bed

Evolution and Function — Historically, railroad freight cars were overdesigned to minimize in-service fatigue failures and high safety factors were used to keep stress levels in critical locations at a minimum. Combined with competition from the trucking industry, increased fuel cost made many of these heavy designs uneconomical; the resulting optimization effort created a need for accelerated fatigue testing. While small-scale material and component testing can very precisely and directly address the science of mechanical fatigue, full-scale testing is often used to verify analyses, address manufacturing issues, test component effects and provide safety from fatigue failures not covered by standard specifications. The Simuloader was built as a facility where multiaxial fatigue tests of entire railcars could be performed at accelerated rates [*1*].

General Description — The Simuloader is a computer-controlled servohydraulic test bed built for the full-scale multiaxial fatigue and vibration testing of railcar bodies. This machine can excite a test car with input profiles and waveforms that are representative of actual railroad service. Train forces are input at the ends of a test vehicle, while suspension displacements are reproduced underneath the vehicle. A side-view of a tank car on the Simuloader is depicted in Figure 1.

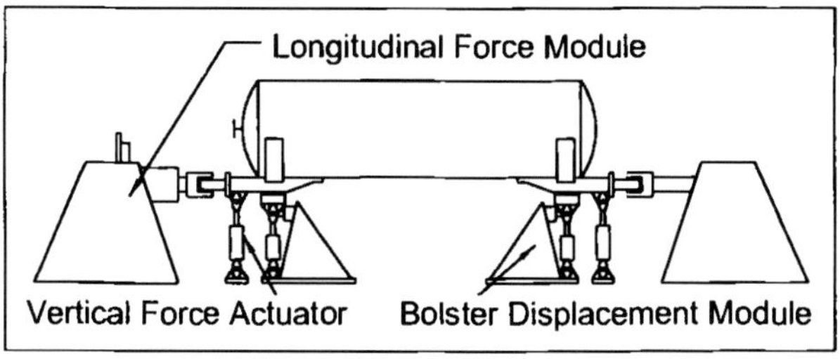

Figure 1 — *Schematic of Tank Car on Simuloader*

This railcar test bed is at the Transportation Technology Center in Colorado, and includes 13 hydraulic actuators of various capacities (with associated servovalves, accumulators and transducers), a six-pump hydraulic power supply and a console of AC and DC analog controllers. A desktop computer is used with a chassis of signal conditioning amplifiers, digital/analog and analog/digital converters to send input signals to the controllers while collecting response data from up to 64 transducers.

Rail Industry Background and Discussion Objectives

Railcar Terminology — The overall assembly of a North American freight car is surprisingly simple. Linkages called couplers connect adjacent railcars to form a train. The wheels and suspensions of each vehicle form subassemblies known as trucks.

Above the trucks, resting primarily on the center plates of the truck bolsters, is the car body. A schematic of a bi-level autorack demonstrates this basic assembly in Figure 2.

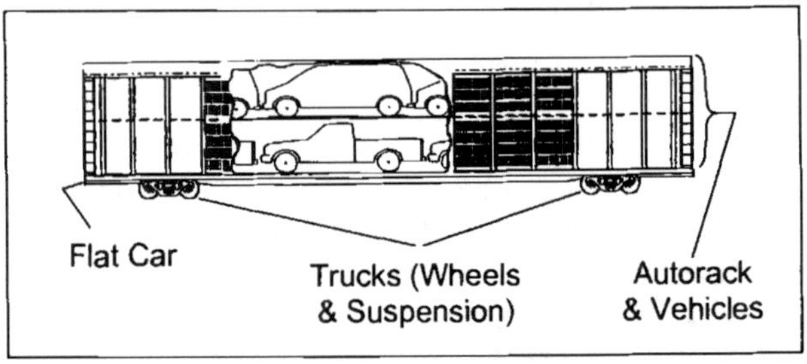

Figure 2 — *Side View of Flat Car with Bi-Level Autorack*

In this case, a two-level superstructure (used to carry automobiles) is attached to and above the platform spanning the two trucks. This autorack superstructure is welded to the flat car, but the car body as a unit is bound to the trucks by gravity only.

A freight car truck assembly is held together vertically by gravity as well. The wheels are rigidly mounted to solid axles (necessary to carry heavy freight loads); two side frames connect the ends of these axles. These side frames carry spring nests and other suspension components (damping elements, etc.), and the truck bolster spans the lateral distance between them. The railcar body rests nominally on the circular center bowl at bolster mid-span, and the side bearings provide roll stability during transit (they can be either pre-loaded or set with a vertical gap to the car body). The basic components of a truck are called out in the isometric view of Figure 3.

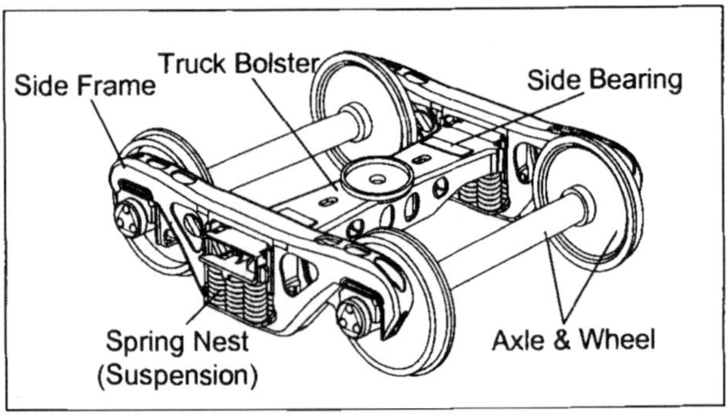

Figure 3 — *Isometric View of Railcar Truck*

Structure of Discussion — Full-scale accelerated fatigue tests are performed by the rail industry for many reasons. In anticipation of the formal adoption of an industry-wide damage tolerance program for tank cars, full-scale experiments were performed to help calibrate and validate the finite element and fracture mechanics models used [*2*].

Several builders, in an attempt to uncover production-related fatigue issues not apparent in early analyses, have subjected full-scale prototypes to accelerated tests. More recently, older designs have been tested to prove robustness and worthiness for extensions to mandatory retirement regulations within the industry.

The discussion presented here is not focused on specific test motivations, but rather on an explanation of typical methodologies used to collect the necessary information and perform these accelerated tests. Examples are drawn from several different past experiments, though results quoted are representative only. The stress analysis of a candidate test car is discussed, followed by the characterization of the service environment. The paper then culminates with the development of machine inputs to animate the test vehicle and system tuning techniques.

Finite Element Stress Analysis of Test Car

Stress Analysis Motivation and Example Project

Fatigue-Significant Thresholds — An underlying assumption in accelerated-rate fatigue testing is that a large percentage of vehicle service data consists of load levels that do not produce fatigue-significant structural stresses. Because the majority of service loads do not propagate fatigue damage and are therefore omitted from the load spectra applied to test vehicles, the effective rates of laboratory simulations are much faster than real time as measured by a clock. Using this concept, the effects of 30 years of railcar service can typically be simulated within a few months. To achieve test rates like this in a fatigue test program, an initial step involves the determination of stress thresholds, or truncation levels, below which events can be removed from laboratory load spectra. This identification of fatigue-significant events has been aided in recent years through the advance of finite element analysis (FEA).

Example FEA Parameters — As a demonstration, a linear finite element stress analysis of a bi-level autorack is used to perform fatigue calculations that subsequently guide laboratory input truncation decisions. Developed from engineering drawings of a bi-level autorack, this model used over 20 000 elements with 18 unique plate thicknesses, four unique beam properties and two distinct mass elements. Basic ASTM A36 steel properties were assumed for all car body structure and non-structural mass elements were added at discrete positions to reflect the weight of brake rigging, miscellaneous equipment and automobiles. Assumptions were made to model corrugated sheet and expanded metal with simple plate elements. Finally, welded connections were modeled as joints between plate or beam elements; no special consideration was given to weld geometry or sizing during modeling.

Load Cases and Constraints — The FEA work focused on the effects of longitudinal coupler force (LCF) from adjacent vehicles, vertical center plate loads, laterally-induced car body roll and car body twist motions; both static and dynamic loadings were investigated. These and a few additional basic rigid- and flexible-body modes of vibration are depicted in Figure 4.

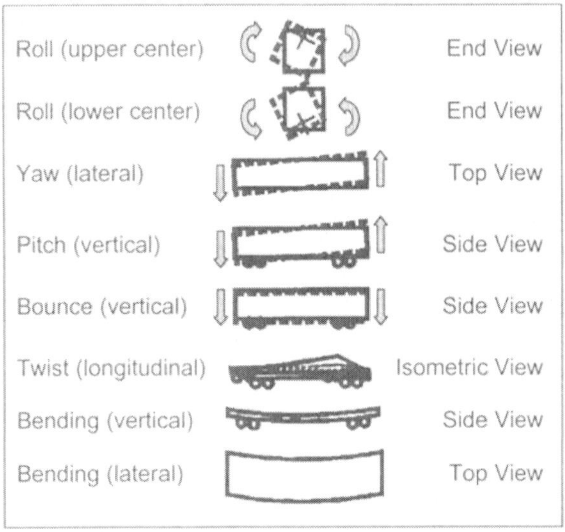

Figure 4 — *Basic Railcar Rigid and Flexible Body Modes*

This railcar model was subjected to LCF, center plate, roll and car body twist loads to simulate both static and dynamic events. To simulate constant-velocity quasi-static train forces, "static" LCF was applied to a quarter model as nodal loads acting at the coupler restraints; symmetrical constraints were used to complete the model. To approximate yard impacts and in-train accelerations between vehicles, "dynamic" LCF was modeled as a longitudinal acceleration acting against longitudinal constraints placed at the coupler restraints (only compressive impacts were simulated). Vertical center plate loads were analyzed with the application of a downward acceleration to the model constrained at the center bowl. Car body roll loads were created with a lateral acceleration applied to all elements, creating a rolling moment reacted at the car center bowl and side bearings. A downward acceleration on the car, while constrained at diagonally opposed side bearings, produced stresses related to car body twist.

Complex Stress States — The model was used to calculate parent metal stress levels within the railcar for the load cases discussed. Because the structure of the railcar acts as one elastic body, stress sensitivities (i.e., the ratio of resultant stress to applied load) were calculated through scaling FEA results to reflect loads greater than or less than those modeled, within the elastic limits of the material. And because the principle of superposition allows different load cases to be summed, the model was also used to investigate the effects of simultaneous load cases. Together, scaling and superposition allowed a limited number of modeled load cases to represent most combinations and magnitudes of the more common service loads.

Several authors have proposed various methods of summing complex stress states into single values for fatigue calculations [*3*]; one method is proposed here. For each modeled load case, static biaxial stress states were recorded for individual elements as two axial and one shear stress component. To examine a cycle between two distinct load states, a mean (offset from zero) and range (peak to peak magnitude) was calculated for each of these stress state components. Using plane stress and octahedral shear (von Mises) principles [*4*], these component means and ranges were then combined into a single non-directional cyclic stress mean and range. This was done for the key model elements in several load case combinations and is demonstrated here in principle. Biaxial (plane stress) principal and von Mises stresses are defined as:

$$\sigma_1,\sigma_2 = (\sigma_x+\sigma_y)/2 \pm [((\sigma_x-\sigma_y)/2)^2+\tau_{xy}^2]^{1/2} \qquad (1)$$

$$\sigma_{vM} = [((\sigma_1-\sigma_2)^2+\sigma_2^2+\sigma_1^2)/2]^{1/2} \qquad (2)$$

where

σ_1 & σ_2 = principal stress magnitudes (e.g., Pa),
σ_x, σ_y, & τ_{xy} = element stress in x- and y-direction and xy-plane shear stress, and
σ_{vM} = von Mises effective stress, based on octahedral shear stress yield criterion.

Substituting equation (1) into (2), von Mises can be found directly for biaxial stress states. With further substitution, two biaxial stress states (i.e., σ_{x1}, σ_{y1}, τ_{xy1}, σ_{x2}, σ_{y2}, and τ_{xy2}) are combined into a single non-directional stress cycle (compression or tension is assumed based on the average principal stresses in both directions) as follows:

$$\sigma_{vM\text{-}mean} = (sign)[(\sigma_{x\text{-}mean}+\sigma_{y\text{-}mean})^2/4+3(\sigma_{x\text{-}mean}-\sigma_{y\text{-}mean})^2/4+3\tau_{xy\text{-}mean}^2]^{1/2} \qquad (3)$$

$$\sigma_{vM\text{-}range} = [(\sigma_{x\text{-}range}+\sigma_{y\text{-}range})^2/4+3(\sigma_{x\text{-}range}-\sigma_{y\text{-}range})^2/4+3\tau_{xy\text{-}range}^2]^{1/2} \qquad (4)$$

where

$\sigma_{vM\text{-}mean}$ = mean (zero offset) of elemental von Mises summed cycle (e.g., Pa),
$\sigma_{vM\text{-}range}$ = range (peak to peak) of elemental von Mises summed stress cycle,
sign = arithmetic sign of the sum of stress component averages ($\sigma_{x\text{-}mean}+\sigma_{y\text{-}mean}$),
$\sigma_{x\text{-}mean}$, $\sigma_{y\text{-}mean}$, & $\tau_{xy\text{-}mean}$ = $(\sigma_{x1}+\sigma_{x2})/2$, etc. (component stress averages), and
$\sigma_{x\text{-}range}$, $\sigma_{y\text{-}range}$, & $\tau_{xy\text{-}range}$ = $(\sigma_{x1}-\sigma_{x2})$, etc. (component stress differences).

Using this, if a parent metal element yielded the following approximate FEA results:

- load case 1 ⇒ σ_{x1} = -46 MPa σ_{y1} = 13 MPa τ_{xy1} = -31 MPa
- load case 2 ⇒ σ_{x2} = 70 MPa σ_{y2} = 8 MPa τ_{xy2} = 43 MPa

The resulting calculated stress cycle between these load cases would be: $\sigma_{vM\text{-}mean}$ = 15.4 MPa and $\sigma_{vM\text{-}range}$ = 175 MPa (i.e., σ_{max} = 103 MPa and σ_{min} = -71.9 MPa).

Note that when von Mises criteria are used, principal stress directions are lost and subsequent fatigue calculations are therefore performed without regard to the stress tensors working specifically against geometric concentrations of interest (e.g., the vector component of a multiaxial load state displacing material in a direction normal to a critical weld toe). While this assumption has proven both conservative and efficient for parent metal stress calculations in the development of full-scale vehicle drive files, it does not provide the necessary accuracy for more detailed and component-specific analytical fatigue predictions. For this latter reason, fatigue calculations in critical weld areas were performed with a different approach (described in the next section).

Sample Fatigue Analysis and Weld Considerations

Parent Metal Fatigue — After selected stress analyses were performed for cyclic loads oscillating between different load cases, fatigue endurance limits were estimated based on guidelines from both the American Welding Society (AWS) and the Association of American Railroads (AAR) [*5-7*]. Modeled load magnitudes were then increased until threshold stress cycles were found (equal to the fatigue limits) for critical areas on the car. This resulted in the minimum dynamic loads that could initiate parent metal (assumed to be flame-cut steel) fatigue cracking. The stress threshold beyond which fatigue damage will propagate for a specific metal type and geometry is defined in the guidelines and can be compared to FEA bounce results with:

$$2aS \leq \sigma_{\text{threshold}} \quad (5)$$

where

$\sigma_{\text{threshold}}$ = stress range (peak to peak) below material endurance limit (e.g., Pa),
a = magnitude (zero to peak) of dynamic bounce acceleration (e.g., m/s^2), and
S = material/geometry sensitivity to vertical bounce acceleration (e.g., Pa-s^2/m).

Using this, if an AWS-defined allowable stress range ($\sigma_{\text{threshold}}$) is 165 MPa and the material/geometry sensitivity (S) to vertical bounce is 87 MPa/G (where G represents the acceleration of gravity, or 9.81 m/s^2), the allowable dynamic bounce magnitude (a) is 0.95×G. In this case, parent metal will accumulate fatigue damage from cycles exceeding 1.0×G ± 0.95×G (e.g., a transient vertical bump centered around the static load of gravity) and service cycles below this can be excluded from laboratory testing.

Weld Stress Concentrations — FEA welds were defined as a series of nodes at the intersection of two or more elements. Concentration factors for critical welds were used to scale parent metal stress levels calculated at their respective model nodes. This scaling included corrections for weld axis orientation, weld throat size, weld geometry, and weld type [*8*]. The first step in applying concentration factors was to rotate the applicable parent metal stress tensors from local (nodal) coordinate systems to angles aligned with weld throat axes, using the following:

$$\sigma^*_x = (\sigma_x+\sigma_y)/2+[(\sigma_x-\sigma_y)/2]\cos(2\phi)+\tau_{xy}\sin(2\phi) \quad (6)$$

$$\sigma^*_y = (\sigma_x+\sigma_y)/2+[(\sigma_x-\sigma_y)/2]\cos[2(\phi+\pi/2)]+\tau_{xy}\sin[2(\phi+\pi/2)] \quad (7)$$

$$\tau^*_{xy} = -[(\sigma_x-\sigma_y)/2]\sin(2\phi)+\tau_{xy}\cos(2\phi) \quad (8)$$

where

ϕ = angle of rotation necessary to align coordinate systems (e.g., rad),
σ_x, σ_y, & τ_{xy} = axial and shear stresses in nodal coordinate system (e.g., Pa), and
σ^*_x, σ^*_y, & τ^*_{xy} = stresses in rotated (weld throat) coordinate system.

Axial stress in a weld was then found through the scaling of parent metal axial stresses by the ratio of parent metal area to weld area. Per unit width, they were scaled by the ratio of parent metal thickness to effective weld throat (defined for different weld types and qualities by AWS guidelines) as follows:

$$\sigma_{axial,weld} = \sigma_{axial,p\text{-}metal}(h_{p\text{-}metal}/h_{weld}) \quad (9)$$

where

$\sigma_{axial,weld}$ = axial weld stress (e.g., Pa),
$\sigma_{axial,p\text{-}metal}$ = axial parent metal stress (*F*/*A*, etc.),
$h_{p\text{-}metal}$ = parent metal thickness (e.g., m), and
h_{weld} = effective weld throat thickness (as defined by AWS).

Weld bending stress was calculated in a similar fashion, using a section modulus ratio:

$$\sigma_{bending,weld} = \sigma_{bending,p\text{-}metal}(Z_{p\text{-}metal}/Z_{weld}) \quad (10)$$

where

$\sigma_{bending,weld}$ = weld bending stress (e.g., Pa),
$\sigma_{bending,p\text{-}metal}$ = parent metal bending stress (*Mc*/*I*, etc.),
$Z_{p\text{-}metal}$ = parent metal geometry section modulus (*I*/*c*), and
Z_{weld} = weld metal geometry section modulus.

For these calculations, it was assumed that no in-plane bending occurred down the length of the weld bead within the individual elements. Out-of-plane bending was identified as the difference between top-of-element and bottom-of-element axial stresses perpendicular to the weld. The averages of these top and bottom values defined axial stresses along those perpendicular axes.

Similar to the computation process used for parent metal, cyclic stress means and ranges were found for each critical weld and load case combination. Based on the type, quality and load path through the weld, endurance limits were read from either AWS plots [*6*] or AAR Modified Goodman diagrams [*7*], and fatigue analyses were performed for cyclic loads in welds. Values used from AAR and AWS data were estimated based on common welding practices for the railcar and welds characterized, with additional factors of safety to cover reduced material area and stress risers typically associated with welds. FEA-based weld fatigue predictions like these often overestimate stress levels and can suggest conservative truncation levels (i.e., lower stress threshold values above which cycles are assumed to be fatigue significant). Of course, the impact of an overestimated stress on a truncation level is also dependent upon the fatigue model in use (e.g., stress-life, strain-life, crack growth) and all of the other input parameters to that model (discussed further in the next section).

FEA Validation and Laboratory Truncation Decisions

Model/Field Correlation — The stress analysis was used to determine sensitivity of the railcar to selected load cases. An additional step is required, however, as what is typically measured in service is not typically output by stress analysis. For field data truncation decisions, analysis output was normalized to quantities measured in field data collection efforts. Modeled and measured bounce was correlated through bolster acceleration rather than force. Likewise, car body roll was quantified through side bearing load and twist through side bearing displacement. Sample autorack sensitivities to common vibration modes are shown in Table 1.

Table 1 — *Car Body Sensitivity to Fatigue Loading*

Load Case	Sensitivity
LCF	0.050 MPa/kN
Bounce	1.35 MPa•s^2/m (13.2 MPa/G)
Roll	0.61 Mpa/kN
Twist	17.9 MPa/mm

Because load environment acquisition often employs many strain measurements, model validations can be performed through comparisons of modeled region sensitivities to those of strain gages (under applied structural loads). In railcar testing, this comparison has often been used to catch errors in both models and measurement strategies. In the case of the example autorack, the model agreed favorably with measured stress sensitivities. However, in a previous tank car modeling effort, it was discovered that certain strain measurements were not oriented ideally and that some modeled stress concentration parameters had to be revisited.

Truncation Threshold Values — Using the methods outlined here, some representative fatigue-damage thresholds and bi-level autorack response sensitivities to car body vibration modes were used to calculate the example truncation recommendations displayed in Table 2.

Table 2 — *Truncation Levels under Fatigue Loading*

Load Case	Parent Metal Threshold	Average Weldment Threshold
LCF	1116 kN	588 kN
Bounce	± 7.74 m/s^2 (0.79×G)	± 2.84 m/s^2 (0.29×G)
Rock	248 kN	93 kN
Twist	8.2 mm	3.6 mm

Combined Load Cases — Vertical and lateral railcar operating modes usually occur independently of one another, but simultaneously with LCF (discussed more later). LCF and bounce, the most commonly measured concurrent loads, were analyzed for combined fatigue effects. In the model, all loads were combined with the static load due to gravity (i.e., a combined LCF of 200 kN and a dynamic vertical load of 3.9 m/s^2 implied simultaneous cyclic loading of ± 200 kN longitudinally and 9.8 ± 3.9 m/s^2 vertically). Again, load magnitudes were increased until response stresses were equal to material endurance limits (to find thresholds).

Combined LCF and vertical load truncation levels are depicted in Figure 5. Fatigue-significant combinations appear in the shaded region in the upper right. Note that while an applied LCF of 350 kN by itself may not cause damage, it can if combined with a dynamic vertical event with a magnitude exceeding about 2.0 m/s^2.

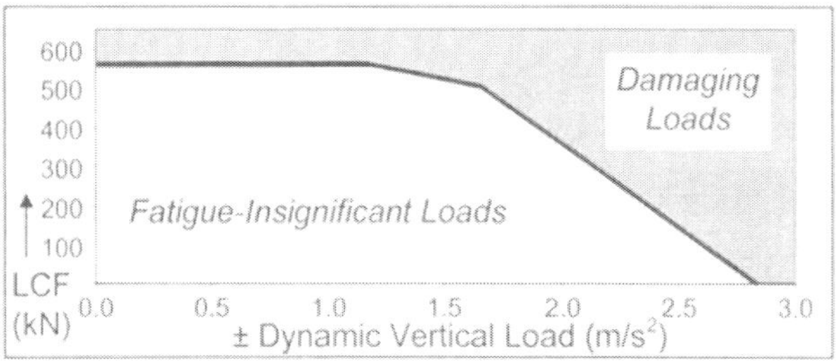

Figure 5 — *Truncation Levels for Combined Loads*

Stress Analysis Summary — Other methods and enhancements certainly exist for determining truncation levels from FEA. This one is presented because it can be used to quickly relate model to field data, while incorporating AWS or AAR fatigue guidelines that are conservative enough to allow universal and rough application during design. The primary advantage of an FEA model is the detailed stress contours and variations within welds that it computes which are not readily determined by classical analytical methods. Other analyses presented help ensure that all fatigue-significant service loads are applied to a test vehicle, without spending valuable simulation time on the replay of non-damaging and insignificant cycles.

Acquisition and Characterization of Service Environment

Measurement of Service Inputs

Machine/Vehicle Interface — In service, vertical and lateral motions of the railcar wheels are transferred to the truck bolster through the suspension, and from there up to the car body. In the case of accelerated railcar fatigue simulations, a direct connection between the test bed and the body of the test car is preferable to one between the machine and wheels/axles of the car (i.e., the trucks are removed for the laboratory fatigue simulation). This is primarily because at high test rates, when many large events are closely strung together, suspension damping elements can heat up and behave abnormally; this can have consequences concerning both simulation accuracy and personnel safety. For this reason, the Simuloader interfaces with test vehicles through simulated bolster modules, with servohydraulic actuators replacing the suspension elements normally resident underneath the truck bolster depicted in Figure 6.

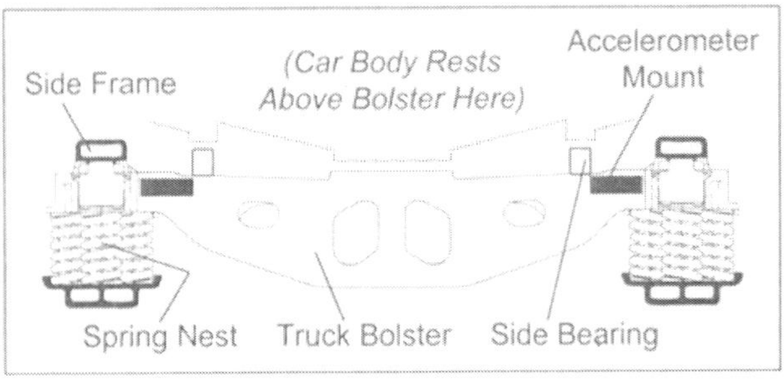

Figure 6 — *Front View of Truck Bolster and Spring Nest*

On the full-scale test stand, actuators create bolster displacements directly underneath the car body. In addition to these displacements, longitudinal coupler forces from in-train measurements are applied to the ends of the car through solid load transfer blocks. Damping normally utilized in longitudinal coupling is removed to improve test control.

Bolster Motions — The displacement control signals used to animate the Simuloader bolster modules are double-integrated from mechanically-damped bolster acceleration time histories measured over selected test routes. A typical service test will include a vertically-oriented accelerometer on each side of the truck bolster and one lateral measurement adjacent to a vertical (see Figure 6). Time histories allow for the retention of phase relationships between the vertical and lateral motions at each end of the car during laboratory simulation (i.e., if the railcar pitches and twists due to a track perturbation in service, it will pitch/twist while traversing the simulated version of that event on the test stand). Because the truck bolster is part of the sprung mass, a measurement of displacement across the suspension would include wheel/axle translations that are absorbed by the suspension and do not affect the car body. Bolster acceleration measurements, however, capture motions relative to ground (Earth), rather than with respect to the wheel/rail interface. Relatively fragile servo-accelerometers are used because they typically have one-tenth the error of more robust piezo-electric (PE) and piezo-resistive (PR) devices; mechanical damping is installed to protect them from high-frequency overload damage. This high level of accuracy is necessary to address the fact that most significant railcar inputs are at low frequencies (0-30 Hz) and might not be good candidates for integration without good measurement resolution. This is demonstrated through a comparison of accelerometers with full-scale ranges of 98.1 m/s^2 ($10 \times G$) and a simple sinusoid represented by:

$$a = -\omega^2 x[\sin(\omega t)] \tag{11}$$

where

a = amplitude of acceleration (e.g., m/s^2),
ω = frequency of periodic waveform (e.g., rad/s^2), and
x = amplitude of displacement (e.g., m).

A pure 2 Hz displacement of ± 5 mm (one of the smaller motions of potential fatigue significance, based on the FEA results presented) would yield an acceleration amplitude of 0.79 m/s^2 ($0.08 \times G$). This is within the uncertainty and cannot be detected reliably by the PE or PR accelerometer (they are typically 1.0% devices), but is several times larger than the error band of a comparable servo device (typically a 0.1% device).

To date, the most common forms of mechanical isolation used to protect servo-accelerometers from high-frequency accelerations have been various grades of foam rubber, applied between the transducers and the mounting surfaces. An example of a material that has been used has an amplification factor (or gain) as in Figure 7.

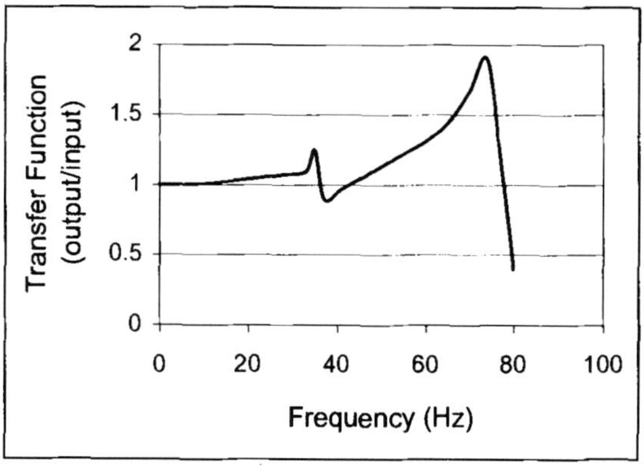

Figure 7 — *Gain of Typical Isolation Material*

Transfer functions like this can be obtained through impact hammer or sine sweep testing; a good reference concerning the details of these types of characterization tests was written by Ewins [*9*]. In this example, the gain of the foam increases to about 1.9, before rolling off at about 80 Hz. This was acceptable for its application because:

- physically damaging accelerations above a frequency of 80 Hz would be mechanically damped, and would not reach a servo accelerometer
- vibration energy occurring below 80 Hz was not expected to be at damaging levels for the transducers, even if amplified by a gain up to 1.9
- collected data would be digitally low-pass filtered (post-test process) at 30 Hz, thereby removing most amplified portions of the spectra

In this example, however, a temperature dependence was discovered; that is, the transfer function of this material changed with temperature. It was therefore concluded that it was best used with ambient temperatures of 50 to 100 °F, when the characteristics listed above were stable and repeatable (summertime test).

Although foam comes with the difficulty that its stiffness and damping coefficients are inextricably linked (i.e., one cannot be varied without affecting the other), it is relatively inexpensive and easy to implement. For proper utilization, a mechanical isolation material requires characterization prior to testing to ensure that the transfer function will work with the intended application. Like an idealized filter, an ideal sample of this isolation material would have a transfer function characterized by a magnitude of one, linear phase within the pass-band and a steep roll-off to zero gain; it would also be totally insensitive to temperature. In the case of railcar bolster measurements, this ideal foam rubber might allow all 0-30 Hz vibrations through without any amplification or attenuation, and stop all other vibration frequencies from physically reaching the transducer. This ideal sample does not seem to exist.

Coupler Forces — Past work has demonstrated that while vertical and lateral suspension displacements originate primarily from track geometry and localized irregularities, longitudinal coupler forces are more closely tied to operator train handling and overall geographic terrain. For this reason, coupler forces and bolster motions have been deemed incoherent (i.e., they are not linearly related, with the possible exception of yard impact situations) and are often measured independently without regard to their phase relationships. The force control signal used to drive the longitudinal actuator of the Simuloader is usually derived from rainflow cycle counted data [*10*] taken from a force-calibrated strain-gaged coupler and stored separately from the bolster time histories. Because rainflow data takes comparatively less computer memory, a smaller system can be used to collect data from longer tests that span many operator/route combinations and capture more statistical outliers (extreme loads). This has enabled the collection of larger and more representative field samples, which is necessary to capture the wide scatter of coupler force data that predominantly drives car body fatigue issues (due to operator dependance and increased design sensitivity, compared to that of bolster motions). North American railroads have tested several car body designs using rainflow cycle counting techniques [*11*], and have compiled an extensive database of the resulting histograms [*7*].

Another consideration in the instrumentation of a test vehicle is that of vertical coupler force (VCF). VCF can come from sources such as grade crossings where there is a sudden change in track stiffness, and as force components of longitudinal impacts where the load application (through the coupler) is not at the same height as the railcar center of gravity, thereby creating a vertical moment arm. These forces are not often significant to through-sill designs, where the center sill extends the full length of the car body. Through-sill flat cars (see Figure 2), for instance, have a lower center of gravity (partially due to the through-sill itself), which lessens this moment arm during a longitudinal force application. VCF is, however, a significant damage source for stub-sill designs like tank cars (see Figure 1), where short draft sills are attached to each end of the car body. Given the same impact scenario, the vertical distance between the coupler and the center of gravity is greater for the tank car, which can result in higher vertical forces that are sometimes reacted in more fatigue-sensitive welded areas. VCF measurements are usually implemented in full-scale laboratory simulations for stub-sill railcar designs, but are not necessarily considered for cars built with full-length sills.

Measurement of Car Body Responses

Critical Region Strains — When older railcar designs are candidates for accelerated fatigue testing, as in life-extension programs where mandated retirements are under question or damage tolerance programs are being initiated [*12*], local car body strain measurements are often guided by common field failures and past experience. Newer prototype designs have often relied on FEA to supplement this experience in the guidance of strain measurement locations and orientations. A critical region strain gage layout recently used in an industry test of a tank car is shown in Figure 8.

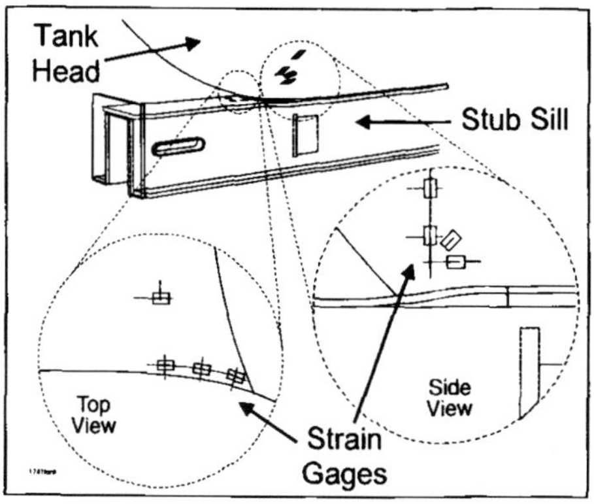

Figure 8 — *Schematic of Tank Strain Measurements*

In this example, researchers were interested in principal stress values and gradients near a flaw that was created with a jeweler saw at a weld joining the tank and stub sill of a tank car. The strain measurements taken during the fatigue simulation were used to validate prior stress analyses and observe changes in overall structural stiffness [1]. In general, railcar instrumentation is focused on regions where high stress gradients and concentrations are expected; areas commonly referred to as critical regions.

Ride Quality Accelerations — Triaxial (longitudinal, vertical and lateral) car body accelerations are measured to help quantify the relative severity of a service test; this helps guide decisions regarding when to stop field data collection, through comparisons to both industry and private databases from previous testing. An example of these comparisons is shown in Figure 9.

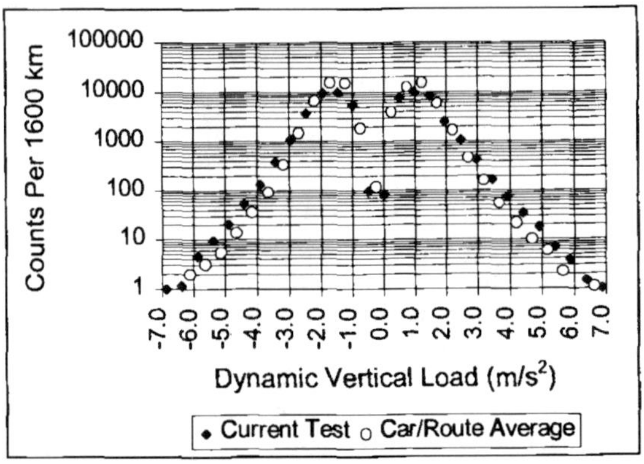

Figure 9 — *Vertical Ride Quality Comparison*

This plot demonstrates that the current service test has captured fewer low-magnitude vertical events than what is average for the specific car and route, but that it has captured more of the high-magnitude event types. Peak-valley cycle counted histograms can be used to help quantify specific service environments. If test parameters (mounting, filter type, sample rate, etc.) are similar to those from previous efforts, there are many sources of industry data available for comparisons like this.

A statistical technique of determining when to end a data collection effort was put forward by Campodonico [13]. Assuming the measured quantity is normally distributed, the variance of the observations can be tracked as a function of the number of observations. If this is done, a normal distribution will develop with a mean and variance. When the shape of this resulting variance distribution stops changing (i.e., the variance of the variance distribution stops changing), the test is over. Whether statistical techniques or past data are used, it is difficult to assess necessary test length a priori; both techniques require the collection of data before they are useful.

Dynamic Operating Modes — In addition to the measurement of specific car body inputs linked to machine requirements and responses linked to stress/fatigue analyses, other instrumentation is used to quantify some basic modes of vibration. These measurements allow for better system tuning and verification that events on the

Simuloader represent events in the field. Usually, the modes characterized include pitch, bounce, twist and roll (see Figure 4).

Like couplers, truck bolsters are often turned into makeshift transducers through shear and bending strain gage circuits; bolster and side bearing vertical loads are then measured to help quantify the body modes listed above. Some gage layout issues that surface in this type of work are with consistency (sensitivity changes due to load path changes), raw output level (needed for a good transducer), linearity in the overall bridge and physical crosstalk between measurements. Bridge types (e.g., shear or bending) and precise gage locations can vary greatly between designs; stress analysis can prevent certain pitfalls, but a degree of trial and error is always expected due to the many differences between the design drawings/tolerances and the actual castings.

Service Test Program Logistics

Sample Rates and Test Durations — Because both the rigid (pitch, roll, etc.) and flexible body vibration modes (twist, bending, etc.) of most railcars are excited below 30 Hz, this is a common analog filter cut-off frequency applied in service testing. To avoid signal aliasing, data acquisition sample rates are set at ten times the filter rates (300 Hz in this case) although recent work in up and down sampling [*14*] may lower these bounds. To ensure consistent comparisons, the filter and sample rates used in laboratory simulations typically follow those used in the field.

As mentioned previously, two different modes of data collection are employed. Bolster motions are collected in the form of continuous parallel time histories from each end of the test vehicle and coupler forces are collected as rainflow cycle counted histograms from one or both ends of the car. Because rainflow data requires less storage space, the coupler data collection effort often will cover approximately 25 000 km, while the bolster test effort may stop after only 3500 km. With this compromise, better statistical surveys are obtained of car body input due to operator train handling (coupler force), but the phase interrelationships between each of the six bolster accelerations (vertical and lateral) are retained for the laboratory simulation.

Train Makeup and Routing — The goal of most vibration/fatigue environment testing is to capture a set of service data that are conservative enough in nature to assure design safety, but not so severe that new car designs based on that data are required to be unreasonably overbuilt. Both empty and loaded operating regimes (autorack with/without automobiles, tank car with/without fluid, etc.) are addressed in the planning of a service test, as each one may excite different modes of fatigue-damaging vibrations that need consideration in the laboratory. In addition to lading (freight type), car position within the consist and overall train length are important in the makeup of a service test, as LCF histograms can be increasingly severe toward the rear of a long train. Finally, the routes, track classes and speeds that a test car traverses need to include a good mix of the expected service of that design (e.g., railroads in the eastern United States tend to have more curves that trains travel slower through, while trains out west tend to travel faster). In short, the aim of a service test is to capture a cross section of service that is representative of what is typical for the railcar tested.

Drive File Development and System Tuning Techniques

Conversion of Service Data to Test Bed Control Signals

Bolster Time Histories — At the conclusion of a service data collection effort, truck bolster acceleration measurements are converted into displacement "drive" signals to actuate the bolster modules of the Simuloader. This data reduction processes begins with the double integration of time histories recorded from the vertical and lateral accelerometers mounted at each end of the car. Again, continuous time histories allow for the retention of relative phasing between bolster animation sources, and accelerations are used to quantify motions relative to ground.

On the surface, the process of double integration seems as simple as applying two consecutive passes of a numerical integration (using the trapezoidal rule, Simpson's approximation, etc.) to discrete data samples. The chief difficulty, however, lies in the selection of a trend removal technique. Trends in service acceleration measurements can come from a variety of sources, including low-frequency elevation changes and superelevation in curves (real accelerations, but not useful for simulations), as well as electrical drift or offsets due to measurement system components (false recordings, not mechanically induced). In addition, trend removal requires the handling of the mathematical constants produced by each integration step. These constants are highlighted in the basic analytical formulation of a double integration by:

$$\iint x''(t)dt = \int[\, x'(t)+C_A]dt = x(t)+C_A t+C_B \qquad (12)$$

where

$x''(t) = d^2x/dt^2 = a(t)$, or acceleration amplitude as a function of time (e.g., m/s^2),
$x'(t) = dx/dt = v(t)$, or amplitude of velocity as a function of time (e.g., m/s),
$x(t)$ = amplitude of displacement as a function of time (e.g., m), and
C_A & C_B = constants defined by boundary conditions (i.e., initial/final values).

In addition to the other sources of low-frequency offsets from zero magnitude, these constants accumulate at successive time steps and result in unstable displacement trends that approach infinite magnitude (limited only by the length of the overall data series). Because laboratory simulation inputs are focused only on relatively finite transient events, these sources of instability must be removed during the integration process.

A simple approach to trend removal is to divide the time history data into finite arrays and subtract the average of each array from itself. While easy to implement, this approach may only eliminate trend components with periods (lengths) close to the window size used to limit the arrays. A slightly more complicated approach uses strict frequency-based filtering. But because several important vibration modes occur at frequencies close to those of specific trend components, this can either eliminate desirable data or allow unwanted trend components to pass through the process. What is needed is an inclusive approach that can address the different sources of trends while minimizing effects on response data related to important body modes.

A trend removal technique that has been developed borrows from proportional, integral, derivative control theory. This technique passes continuous data through what is effectively a trend-following, high-pass, adaptive filter. This dynamic filter has a similar effect on data as that of a strict frequency-based filter, but the effective filter shape (i.e., pass-, transition- and stop-band tolerances and frequencies) is continuously adjusted in response to changes in the data received. The proposed application of this dynamic-filtering concept uses the following trend-removal computations in conjunction with each iteration of the numerical integration:

$$d^*_n = d_n - (K_P d_{(n-1)} + I_{(n-1)}) \tag{13}$$

$$I_n = I_{(n-1)} + K_I d^*_n / s \text{ (alternatively, } I_{(n-1)} = I_{(n-2)} + K_I d^*_{(n-1)}/s) \tag{14}$$

where

d^*_n = corrected data point magnitude (after trend removal filter step),
d_n = original data point magnitude (e.g., output of numerical integration step),
K_P = proportional feedback constant (dimensionless),
K_I = integral feedback constant (dimensionless), and
s = data sample rate (e.g., Hz).

Tuning techniques for adjusting K_P and K_I are beyond scope of this discussion, but a beginner's reference was put together by ExperTune, Inc. [*15*]. This dynamic filter can be tuned based on typical bandwidth and amplitude content of signals under scrutiny.

In a previous controlled track test, a string potentiometer was mounted across the spring nest of a truck (see Figure 3), and accelerometers were applied to the bolster end and side frame mid-span such that they could be summed to gather acceleration data across the same spring nest. A similar lateral configuration of transducers was also applied to the truck. Data from these was collected over a cross section of typical vertical, lateral and combined perturbation types, and then used to calibrate the dynamic filtering process prior to releasing the fully-instrumented railcar for service testing. The acceleration signal of the bolster relative to the side frame was integrated twice and compared to measured bolster displacement relative to the side frame. Through systematic trials of a range of filter constants on several event types expected in service, constants of 0.2 and 2.0 (K_P and K_I, respectively) were selected as an optimal balance over a broad range of typical event frequencies. A comparison of some directly-measured versus double-integrated (using these constants) vertical displacements resulting from railroad twist perturbations is shown in Figure 10.

In this plot, the lighter signal is the string potentiometer measurement and the heavier one is the product of the double integration process with dynamic filtering. While the waveforms do not precisely match, the integrated acceleration amplitudes are representative of the measured displacements during the event. For increased simulation accuracy, cross-axis correction factors can be applied to lateral displacements measured in the field due to car body roll, as those motions should not be applied on a test stand as lateral translations. Simple geometric correction factors can also be applied to vertical field data to account for any lateral differences between the vertical measurements and test stand load application points.

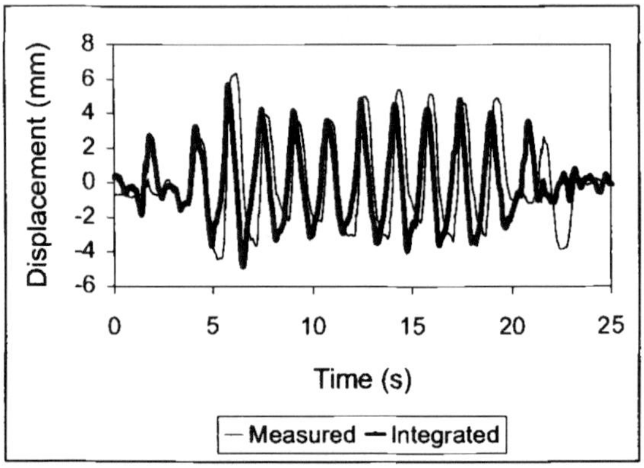

Figure 10 — *Integrated and Measured Displacement*

Coupler Cycle Counts — Also at the conclusion of a service data collection effort, variable-amplitude train forces (stored as rainflow cycle-counted data) are converted into a sinusoidal force drive signal to actuate the longitudinal coupler actuator of the Simuloader. Because multiple passes through a single load schedule (e.g., rainflow histogram) are required for the completion of a fatigue test, the overall cycle application order in an entire test is effectively random regardless of the internal ordering within an individual schedule. And because the relative timing between LCF cycles and bolster motions is not usually retained during service data collection (previous studies indicate that they are largely incoherent), a randomization of coupler load sequence relative to that of bolster motions is acceptable. This latter statement is made with an exception: the few high-magnitude cycles in histograms due to low-speed yard impacts are often separated for solitary application on the test stand without a parallel application of bolster displacements that were collected at higher speeds. When multiple longitudinal load paths are a consideration in a car body, impact testing can be performed on track in conjunction with Simuloader operations, as the machine is not capable of applying forces at high impact frequencies. Cycles within a typical LCF drive signal are applied at frequencies that vary between the controllable limits of the actuator (0-10 Hz).

Drive File Condensation — With the FEA and service data collection efforts complete, the final stage of input development involves piecing together what will become a drive file used to animate the vehicle test bed for a period of time that represents an accelerated quantity of potential fatigue damage. The first step of this final stage is centered around truncation, or removing non-damaging events from the service data and creating accelerated machine inputs for the test vehicle. Truncating the LCF rainflow histogram is done before the coupler sinusoid is created through elimination of cycles in bins below identified stress thresholds. Truncation of bolster displacements is somewhat more involved, however, as six time histories are operated upon (two vertical and one lateral bolster acceleration at each end of the car).

Using key response data channels (critical region strains, ride quality accelerations, etc.) and FEA-based car body sensitivities, the rail service data is scanned for events of bounce, twist, and roll that are considered to be fatigue significant. When potentially-fatiguing zones of operation are isolated, the corresponding bolster motions are extracted for laboratory simulation. At times, zone identification is also based on various summations of the bolster accelerations themselves. A time history of vertical bolster acceleration associated with a grade crossing is plotted in Figure 11; this event is typical of what is selected for laboratory simulation. In this example, the lead truck hit the grade crossing about one second before the trail truck (the truck spacing was about 20.1 m and the speed of travel was 76.5 km/h). For simulation inputs, the following is extracted: some time leading into the event, the transient event itself (time while any portion of the railcar is either approaching or exceeding defined fatigue thresholds) and time following the event as the railcar settles back into normal operation.

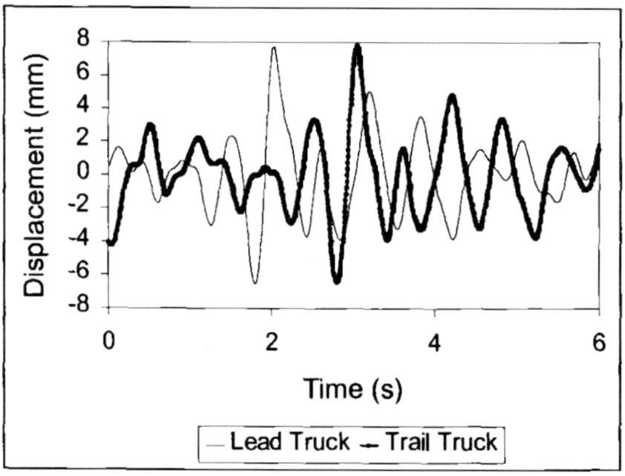

Figure 11 — *Vertical Input for Grade Crossing*

While the pre- and post-event periods used during this form of data extraction must be long enough that neighboring events do not influence each other during simulation, they cannot be so long as to significantly lower overall laboratory acceleration rates. Past experience in controlled track testing has led to pre- and post-event periods based on test vehicle length and speed; typical time windows have been such that data is retained from a distance equal to half the car length before the location of an event and one car length after. These pre- and post-event periods do not include the actual event length, which is defined as the time while any portion of the railcar, lead or trail, is either approaching or exceeding defined fatigue thresholds.

After significant bolster events have been selected, they are concatenated in series with tight cubic splines (both brief in time and small in magnitude) inserted between them to smooth discontinuities between end and initial states of adjacent events. This is done to safeguard against harmful and dangerous shocks to the machine and test vehicle. At this point, the truncated sinusoid created to actuate the coupler is patched in parallel with the six bolster displacement waveforms. The last hurdle in this part of the process is matching the time scales between accelerated inputs. If a truncated LCF sinusoid representing 9000 km of service is combined with a set of bolster displacements representing 6000 km of service, three repeat passes through the bolster motions are paired with two passes of LCF to make a combined drive file. This composite drive file, representing 18 000 km of fatigue-significant car body service, would then be tuned based upon laboratory vehicle response (described further in the next section) and repeatedly applied to the test car until the expected equivalent lifetime was accumulated. Typical accelerated test rates for railcars have ranged from 5000 to 25 000 km of fatigue-significant service per hour of laboratory simulation.

Servohydraulic Test Bed and Vehicle System Tuning

Vehicle/Machine System — Because drive files for the Simuloader are prepared from directly measured field inputs, differences in how the car body responds to the machine from its response to the trucks can surface. This has to do with several factors, including machine versus truck suspension stiffness differences and phase distortion of individual inputs, due to both analog frequency filtering (used during field data collection) and adaptive trend-removal filtering (used during double integration). While the effects of these differences can be minimized, they are still always present and can affect car body response on the test stand.

To combat these difficulties, several test laboratories in the ground vehicle industry have adopted vehicle response parametric control methodologies. In short, this technology uses vehicle responses to create and correct machine inputs through a frequency-response impedance matrix for the machine/vehicle system, empirically developed in the laboratory. Unfortunately, this has proven difficult to implement in full-scale railcar tests to date, as low-frequency measurement resolution and railcar non-linearities (friction damping in trucks, side bearing spacing, center bowl/plate connection, etc.) have limited the accuracy of the matrix inversions necessary to develop input control signals. Instead, the Simuloader currently requires a somewhat more subjective process to tune the system (close the open loop between the gross inputs and responses) and increase vehicle response accuracy. This process uses a combination of analytical and statistical computations.

Periodic Test Observations — At the start of testing, gains applied to various input levels (and even individual actuator control parameters, at times) are adjusted to increase vehicle response accuracy. Response on the Simuloader is evaluated through response data from service. Service/simulation comparisons of these data are done with statistical distributions, frequency domain spectra, fatigue calculations and raw time histories. To ensure repeatability, data from service are periodically compared to data from the laboratory throughout the test. A sample rainflow cycle count comparison of this nature is depicted in Figure 12.

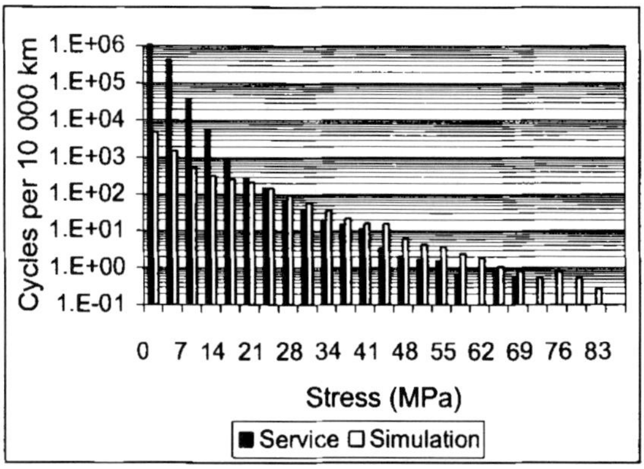

Figure 12 — *Simuloader Tuning Comparison*

In this case, there are fewer low-amplitude stress cycles in the laboratory simulation (a direct result of truncation), but more cycles at the high end than were observed in service. This amounts to a conservative edge on the fatigue damage applied to this specific location in the full-scale test.

As the simulation progresses, periodic damage inspections and physical fatigue crack growth measurements are used to assess design (and model, in some cases) performance. Static car body compliance (stiffness) and dynamic responses are also monitored for change. These latter evaluations can be done with static squeeze tests and frequency sweeps (while monitoring response instrumentation), in addition to the variable-amplitude fatigue inputs.

Summary

The previous discussion was not intended as an explicit statement of all possibilities for full-scale accelerated testing. Rather, it was focused on outlining typical methods and pitfalls in a practical, real-world application of a multi-disciplinary approach for railcars. The stress analysis of a test car, service environment characterization, machine input development and system tuning techniques were discussed with representative examples drawn from previous industry work.

Acknowledgements

The authors would like to acknowledge the following for their financial support and guidance in developing the techniques outlined: the Transportation Technology Center, Inc. (custodian of the Simuloader), the Federal Railroad Administration (owner of the Simuloader), the Association of American Railroads, the Railway Progress Institute. Also acknowledged are the numerous private car builders/owners who have performed full-scale railcar simulations in the past.

References

[1] Smith, K. B., Cardinal, J. W., and McKeighan, P. C., "Stub Sill Tank Car Damage Tolerance Analysis Full-Scale Validation Test," *Proceedings of the International Mechanical Engineering Congress and Exposition,* D. H. Stone, Ed., American Society of Mechanical Engineers, Dallas, TX, 1997.

[2] Cardinal, J. W., "Damage Tolerance Analysis of Tank Car Stub Sill Cracking," Project 06-6965, Southwest Research Institute, San Antonio, TX, 1998.

[3] *Multiaxial Fatigue: Analysis and Experiments, AE-14,* Society of Automotive Engineers, Warrendale, PA, 1989.

[4] Dowling, N. E., *Mechanical Behavior of Materials,* Prentice-Hall Inc., Englewood Cliffs, NJ, 1993.

[5] *Railroad Welding Specification - Cars and Locomotives,* ANSI/AWS D15.1-93, American Welding Society, Miami, FL, 1993.

[6] *Structural Welding Code – Steel, Sixteenth Edition,* ANSI/AWS D1.1-98, American Welding Society, Miami, FL, 1998.

[7] "Fatigue Design of Freight Cars," *Manual of Standards and Recommended Practices, Section C-II, Chapter VII,* Association of American Railroads, Washington, DC, 1988.

[8] Weaver, M. A., "Determination of Weld Loads and Throat Requirements Using Finite Element Analysis with Shell Element Models - A Comparison with Classical Analysis," *Welding Journal,* American Welding Society, Miami, FL, 1999.

[9] Ewins, D. J., *Modal Testing: Theory and Practice,* Research Studies Press Ltd., Somerset, England, 1984.

[10] *Fatigue Design Handbook, AE-10,* Society of Automotive Engineers, Warrendale, PA, 1988.

[11] Sharma, V., and Punwani, S. K., "Freight Equipment Environmental Sampling Test Description and Results," *Proceedings of the Rail Transportation Division Spring Conference,* American Society of Mechanical Engineers, Chicago, IL, 1984.

[12] Stone, D. H., and Cardinal, J. W., "Application of Damage Tolerance Principles to Stub Sill Tank Car Cracking," *Proceedings of the Sixth International Heavy Haul Conference,* International Heavy Haul Association, Cape Town, South Africa, 1997.

[13] Campodonico, S., "Analysis of Vibration Data: When to Stop Collecting," University of Illinois, Urbana-Champaign, IL, 1992.

[14] Vold, H., and Morrow, D., "Compression of time Histories Used for Component Fatigue Evaluation," *International Congress and Exposition, Technical Paper 930403,* Society of Automotive Engineers, Detroit, MI, 1993.

[15] "What is PID – Tutorial," *http://www.expertune.com/tutor.html,* ExperTune Inc., Hubertus, WI, 1999.

Roy L. Hewitt[1]

Real-Time Simulation of a Multi-Channel Moving Load Cell Structural Test

Reference: Hewitt, R. L., "**Real-Time Simulation of a Multi-Channel Moving Load Cell Structural Test,**" *Applications of Automation Technology in Fatigue and Fracture Testing and Analysis: Fourth Volume, ASTM STP 1411,* A. A. Braun, P. C. McKeighan, A. M. Nicholson, and R. D. Lohr, Eds., American Society for Testing and Materials, West Conshohocken, PA, 2002.

Abstract: Full-scale aircraft structural fatigue tests are extremely complex from a control systems viewpoint. There are usually a large number of actuators with significant interactions between them and control is made more difficult because the load cells usually move with the actuators. A linear state space model of a structural test was previously developed as an aid to understand these tests better.

This paper presents a method by which these models could be interfaced with a structural test controller to allow a test engineer or operator to run a virtual test prior to assembling test hardware. An example of a simple cantilever beam with two actuators connected to an analog servo-controller is described. It was possible to simulate the beam using four modes and two actuators with integration step sizes of up to 1 ms. The interaction between actuators could be demonstrated, and the system shows excellent potential as an education tool for new operators. Larger systems can be simulated since the execution time for this example was less than 100 μs. Systems that are more complex could be simulated using a combination of more powerful controller boards

Keywords: aircraft, control, fatigue, full-scale testing, modeling, servo-hydraulic, simulation

Full-scale aircraft structural tests are extremely complex from a control systems viewpoint. There are usually a large number of actuators and the response of any one actuator depends on the control parameters, the actuator dimensions, the servo-valve characteristics, the other actuators, the structure and the loading attachments and reaction fixtures. The situation is further complicated because it is usually necessary to attach the load cells to the moving actuator rod rather than fixing them to the reaction structure as in load frame applications. The disturbances caused by the moving load cell significantly reduce the allowable gain that can be used and make test control more difficult [*1*].

There are also pressures on the test engineer both to set up and complete the test more quickly, in spite of the increasing complexity of the test spectra and loading systems. Thus the need has arisen for tools to help the test engineer better understand the test system and be able to predict the influences of various changes that he can make during the test design phase.

[1] Senior research officer, Institute for Aerospace Research, National Research Council of Canada, Montreal Road, Ottawa, Ontario, Canada, K1A 0R6.

Some earlier work [2] documented the development and experimental verification of a linear state space model of a single channel structural test. It was shown to be sufficiently accurate to allow a greater understanding of these systems and predict the relative stability of different systems. This was extended to a multi-channel configuration [*3*] using modal analysis concepts.

While the state space model in its current form is very useful for understanding the system, it would be more useful for a test operator if it could be interfaced directly with the control system so that the interface was familiar. It could then be used to gain experience in tuning these moving load cell tests and to experience the effects of actuator cross coupling. Ultimately, the test operator or test engineer could then tune the test using all of the tools available within the control system in the same way as for a real test and develop tuning parameters prior to the arrival of hardware.

This paper documents an initial feasibility study using a dSpace[2] real-time controller board connected to an analog servo controller. A description of the hardware and software is provided and results from several simple models are presented.

Real Time Controller System

Hardware

The hardware consisted of a dSpace DS 1102 floating point controller board with a Texas Instruments[3] TMS320C31/40 Mflops processor, 4 ADCs and 4 DACs. The controller board, which is a half-length 16-bit PC/AT slot, was installed in a 25 MHz 386 running Dos 6.2. This computer was used because it was semi-portable and could therefore easily be transported to different tests.

Software

Access to the DS1102 was coordinated via a DSP Device Driver, DSPDRV, while loading and execution of application programs was achieved using the program loader, LD31. Both of these programs were included with the dSpace system and installed on the PC.

Application programs were developed using the Texas Instruments TMS320C compiler / assembler / linker tool set.

Initial Application

A simple example program was developed that performs the simulation of a second order spring-mass-damper system by using Euler integration. It was based on a timer interrupt driven technique in order to obtain a regular sampling period.

A function generator was connected to the input on the real-time controller board and the output was connected to an oscilloscope. The program was then compiled, linked, and loaded to the DS1102. Applying a square wave input yielded the expected response on the oscilloscope.

---

[2] dSpace GmbH, Paderborn, Germany.

[3] Texas Instruments, Houston, Texas.

The Euler integration scheme updated the states using a step size of 100 μs. By progressively reducing the step size until the processor was overloaded, it was found that the minimum step size that could be used was about 10 μs. The calculation time for more complex problems will be much larger than for this simple problem. Furthermore, based on previous simulation work, it was expected that Euler step sizes of less than 10μs would have to be used to obtain stable solutions. It was therefore necessary to develop integration schemes that could take advantage of the linearity of the system if practical real-time solutions were to be obtained.

Integration Methods

Euler Integration

Given the general state space equations

$$\dot{x} = Ax + Bu$$
$$y = Cx + Du$$

the output can be obtained once the differential equation

$$\dot{x} = Ax + Bu$$

is solved. The simplest numerical solution technique is Euler integration where x at any time interval is calculated as

$$x_n = x_{n-1} + \dot{x}_{n-1}\Delta t$$

where

$$\dot{x}_{n-1} = Ax_{n-1} + Bu_{n-1}$$

and Δt is the time increment.

Integration using the Transition Matrix

If x is a single variable, there is an analytical solution for a step response in terms of exponentials. The solution can then be obtained over any time interval where the input is constant, for example, the update rate on a digital controller. A similar, but more complex solution is available when x is a matrix.

It can be shown [4] that the solution at time interval n can be written in terms of the solution at the previous time interval as

$$x_n = \Phi x_{n-1} + \Gamma u_{n-1}$$

where

$$\Phi = e^{A\Delta t}$$

and

$$\Gamma = \int_0^{\Delta t} e^{At} dt B$$

Φ is often referred to as the transition matrix and can be calculated as

$$\Phi = 1 + \Delta t A + \frac{\Delta t^2}{2!} A^2 + \frac{\Delta t^3}{3!} A^3 + \cdots$$

although it is available directly in MATLAB®[4]. It is then a simple matter to show that Γ can be calculated as

$$\Gamma = A^{-1}[\Phi - 1]B$$

Two approximations are involved in a solution obtained by the transition matrix. The first is in the truncation of the infinite series for phi and gamma. However, this can easily be made negligible by the choice of the number of terms unless the step size is so large that precision and round off errors become significant. The second source of approximation error is the discretization of the input. The method assumes that the input is constant over the time step Δt. Therefore, for the example problem above, where the input is fixed, the solution should be accurate for even very large time steps.

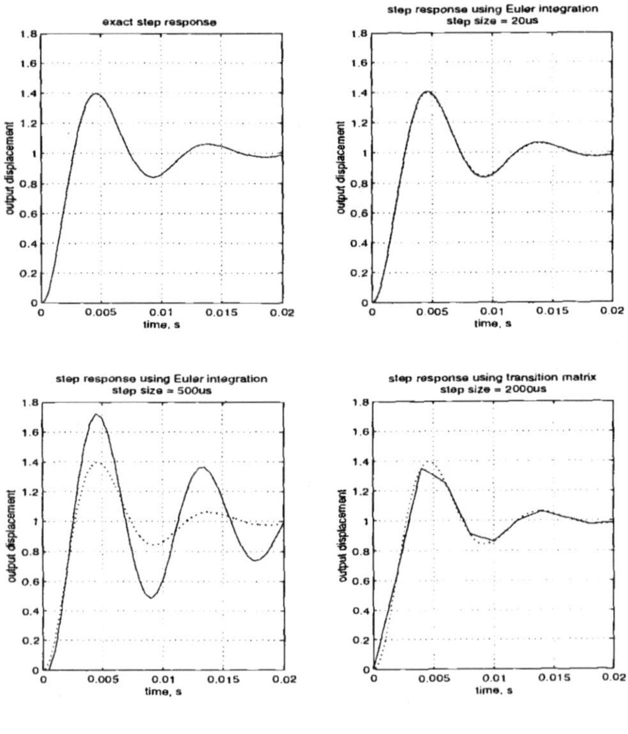

Figure 1 - *Step response using Euler and transition matrix integration.*

[4] MATLAB®, The MathWorks, Inc., Natick, MA.

Evaluation of Transition Matrix Solution

In order to evaluate the integration methods without the complexity of the real-time hardware, a simple MATLAB program was written to perform the calculations that would be performed on the DS1102. Figure 1 compares the exact step response of this linear system with responses calculated using both Euler integration and the transition matrix integration method using various step sizes. The result using Euler integration is accurate with a 20 μs step size. However, if the step size is increased to 500 μs, which is a more realistic time step for real-time solutions of complex systems, the response is incorrect and shows a much less damped system.

The step response using a transition matrix solution with a step size of 2 ms shows some discretization of the response since it is only calculated at each step, but the values at the calculated points are accurate. Thus, for a linear system, where the input is not varying, the transition matrix solution yields an accurate solution for the response even at very large time steps.

Simple Closed Loop System

Open Loop Model

To evaluate the dSpace system in a closed loop situation, a test system consisting of a simple spring/mass structure driven by a servo hydraulic actuator was simulated. The parameters used were for a 5 kN, 15 cm stroke, actuator with a 9.5 l/min servo valve. The spring stiffness was 5.85 MPa and the mass was 4.55 kg.

The first step in the evaluation was to generate the state space matrices for the system. This was accomplished as described in Reference 2 except that only the open loop model is derived as in the block diagram of Figure 2, because loop closure is to be performed using a real controller.

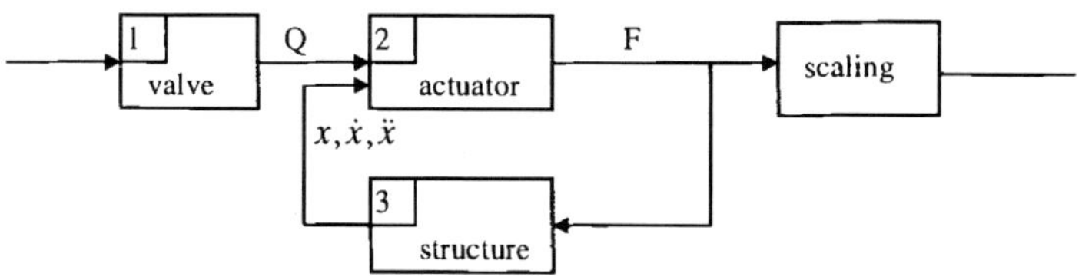

Figure 2 – *Block diagram of structural test system.*

Because it is important to reduce the amount of real-time computation required, the number of states must be kept to the minimum and redundant states should be eliminated. A revised representation of the hydraulic actuator was therefore developed that eliminated one state by using differential pressure rather than the two chamber pressures. Then the actuator model becomes

$$\{\Delta\dot{P}\} = \begin{bmatrix} 0 & 0 \\ 0 & 0 \end{bmatrix} \{\Delta P\} + \begin{bmatrix} \frac{4\beta}{V} & 0 & -\frac{4A\beta}{V} & 0 \\ -\frac{4\beta}{V} & 0 & \frac{4A\beta}{V} & 0 \end{bmatrix} \begin{Bmatrix} Q \\ x \\ \dot{x} \\ \ddot{x} \end{Bmatrix}$$

$$L = [A]\{\Delta P\} + [0 \quad 0 \quad -f_r \quad -m_r] \begin{Bmatrix} Q \\ x \\ \dot{x} \\ \ddot{x} \end{Bmatrix}$$

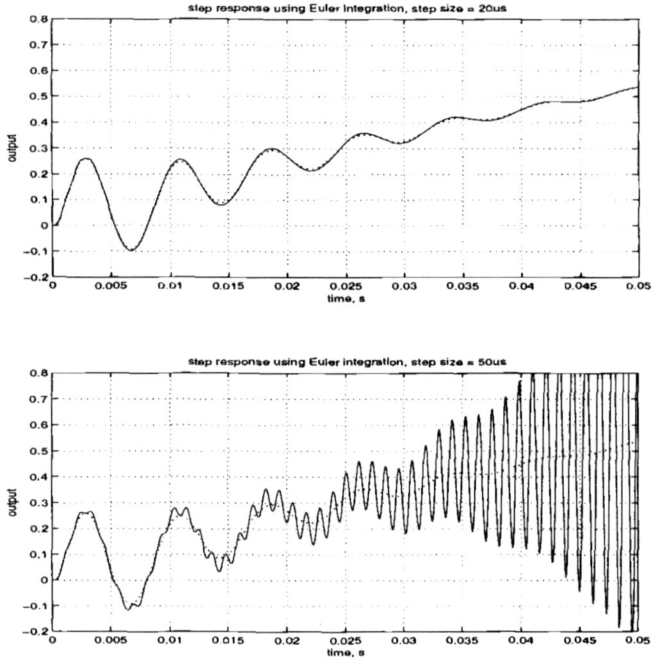

Figure 3 - *Step response of spring system using Euler integration.*

Closed Loop Model

To verify the stability of this system prior to transferring the matrices to the dSpace environment, the closed loop model for a unit negative feedback was generated. Step responses were then calculated using the open loop matrices and the Euler and transition matrix integration schemes to simulate the response that would be expected using the real-time calculations. For these cases, the input was calculated at each time step as the difference between the initial step and the current output, i.e., the error.

Euler Solution

The exact closed loop step response is compared with solutions using Euler integration with step sizes of 20 μs and 50 μs in Figure 3. The response with a step size of 20 μs is reasonably accurate while the one with a step size of 50 μs produces an unstable response.

Transition Matrix Solution

The expression developed above for Γ, i.e.

$$\Gamma = A^{-1}[\Phi - 1]B$$

can only be evaluated for a non-singular matrix A. Because the A matrix for the hydraulic actuator is zero, the combined A matrix is singular for any systems with actuators and an alternate method of calculating Γ is required.

A recursion method for the calculation of Φ is developed in Reference 4 by writing the expression for Φ as

$$\Phi_n = 1 + tA\left(1 + \frac{t}{2}A\left(1 + \frac{t}{3}A\left(1 + \cdots + \frac{t}{n-1}A\left(1 + \frac{t}{n}A\right)\cdots\right)\right)\right)$$

and carrying out the recursive computations

$$\psi_1 = 1 + \frac{t}{n}A$$

$$\psi_2 = 1 + \frac{t}{n-1}A\psi_1$$

$$\psi_3 = 1 + \frac{t}{n-2}A\psi_2$$

$$\cdots\cdots\cdots\cdots\cdots\cdots$$

$$\Phi_n = 1 + tA\psi_{n-1}$$

Now since

$$\Gamma = \int_0^{\Delta t} e^{At}\,dtB$$

then integrating

$$\Gamma = \int_0^{\Delta t}\left(1 + tA + \frac{t^2}{2!}A^2 + \frac{t^3}{3!}A^3 + \cdots\right)dtB$$

yields

$$\Gamma = \left(\Delta t + \frac{\Delta t^2}{2}A + \frac{\Delta t^3}{3!}A^2 + \frac{\Delta t^4}{4!}A^3 + \cdots\right)B$$

which may be written

$$\Gamma_n = \Delta t\left(1+\frac{\Delta t}{2}A\left(1+\frac{\Delta t}{3}A\right)\left(1+\cdots+\frac{\Delta t}{n-1}A\left(1+\frac{\Delta t}{n}A\right)\cdots\right)\right)B$$

and evaluated using the recursive computations

$$\psi_1 = 1+\frac{\Delta t}{n}A$$

$$\psi_2 = 1+\frac{\Delta t}{n-1}A\psi_1$$

$$\psi_3 = 1+\frac{\Delta t}{n-2}A\psi_2$$

....

$$\Gamma_n = t\psi_{n-1}B$$

where only the final computation is different to that for Φ.

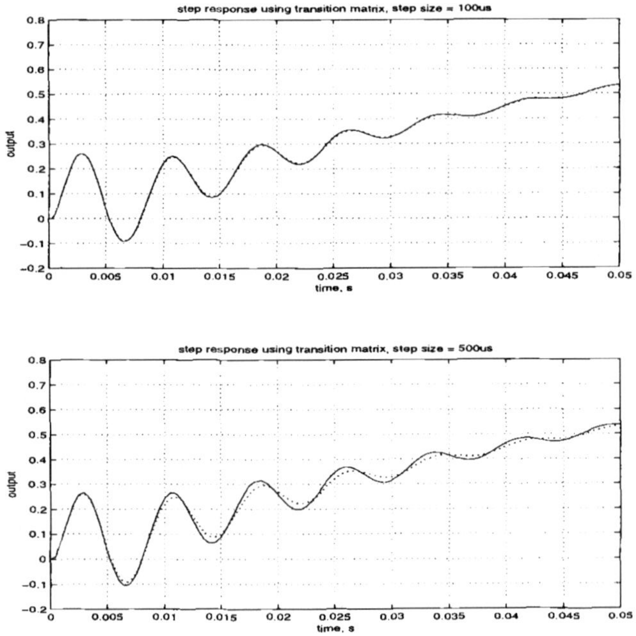

Figure 4 - *Step response of spring system using transition matrix integration.*

As noted above, the transition matrix integration method assumes that the input is constant over the time step Δt. Therefore, for an analog controller, there will be errors introduced by the integration routine because the input (the error signal) would be continually changing in a real test system.

The step responses using transition matrix integration are compared with the exact response for step sizes of 100 and 500μs in Figure 4. The result for a step size of 100 μs is more accurate than the Euler integration with a step size of 20 μs. Even for the large step size, the solution is stable although the system is a little less damped. Because increasing the number of terms in the recursions beyond about 10 had no effect on the result, the difference between the transition matrix solution and the exact solution must be caused solely by the assumption of a constant input. To confirm this, a zero order hold with a sampling time corresponding to the step size was added to the model. Comparisons of the transition matrix solutions and the zero order hold solutions for step sizes of 500 μs and 2 ms are provided in Figure 5. The two solutions are indistinguishable, even with a step size of 2 ms.

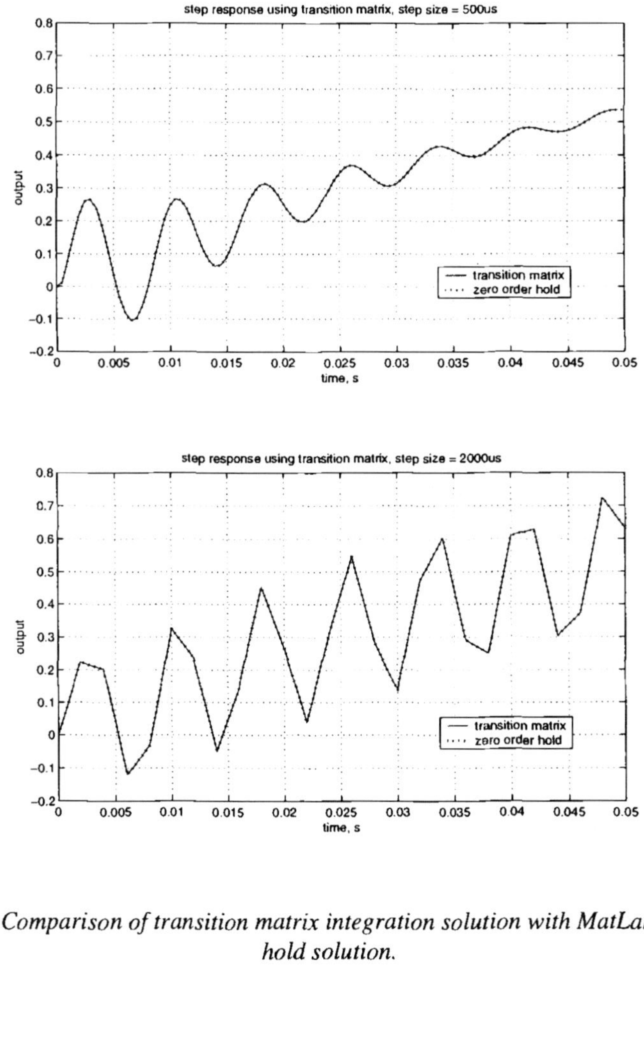

Figure 5 - *Comparison of transition matrix integration solution with MatLab zero order hold solution.*

Real Time Implementation

The basic DSP application program was modified to use the transition matrix solution with six states. Thus the state vector at time step n was calculated as

$$x_n = \Phi x_{n-1} + \Gamma u_{n-1}$$

and the updated output calculated as

$$y_n = C x_n$$

since the D matrix for this system was zero.

The C matrix for the open loop system was copied to the program along with the Φ and Γ matrices for both a 250 µs and a 500 µs step size. While the B and C matrices are the same for the different step sizes, the Φ and Γ matrices are both dependent on step size.

The program was then compiled, linked, and loaded using the option to generate optimized code. This results in a substantial reduction in execution time and is essential for most of these applications.

A function generator was connected to the program input of an analog servo controller. The servo controller error signal was then fed to the input of the real-time controller board. The output from the controller board was connected to the external conditioner input of the servo controller because the output from the DAC on the DS1102 is already a 0-10V signal and further amplification by the servo controller conditioner was not required. It was also connected to one channel of an oscilloscope for display.

With integral gain and reset both set to zero and proportional gain set to one, the response to a square wave input was similar to that observed in Figure 4. The response was sensitive to proportional gain as expected. Addition of integral gain and reset also modified the response, and it was possible to use the system to see the effects of the gain parameters and thus effectively tune the system.

Using a sine wave input, it was possible to see the effects of the various gain controls on the peak error and lag. The system therefore clearly has promise as an educational tool for new operators to learn the effects of system parameters.

Execution Times and Stability

The execution time for the real-time program, found by successively reducing the timer interrupt period, was found to be about 68 µs using standard compilation and about 32 µs using optimized code. Thus, it does appear feasible to simulate a more complex system if reasonably large step sizes can be used.

The stability of the system was dependent on the integration step size. Instability occurred at a gain of about two when using a 500 µs step and at about seven when using a 250 µs step. Smaller step sizes allowed a larger gain before instability occurred.

To investigate how much of the instability was directly attributable to the numerical integration, a numerical simulation was performed for two step sizes using a gain of 10. The result is shown in Figure 6. While the response for a step size of 100 µs is quite stable (and similar to the analytical response), the larger step size produces a response that is very close to instability. However, the transition matrix solution is identical to the zero order hold solution, so the reduction in stability is caused directly by the discretization of the input.

The value of gain required to produce the instability in the numerical prediction was larger than was found using the analog controller. Thus, some of the instability observed in practice must be caused by non-linearities or other effects in the controller.

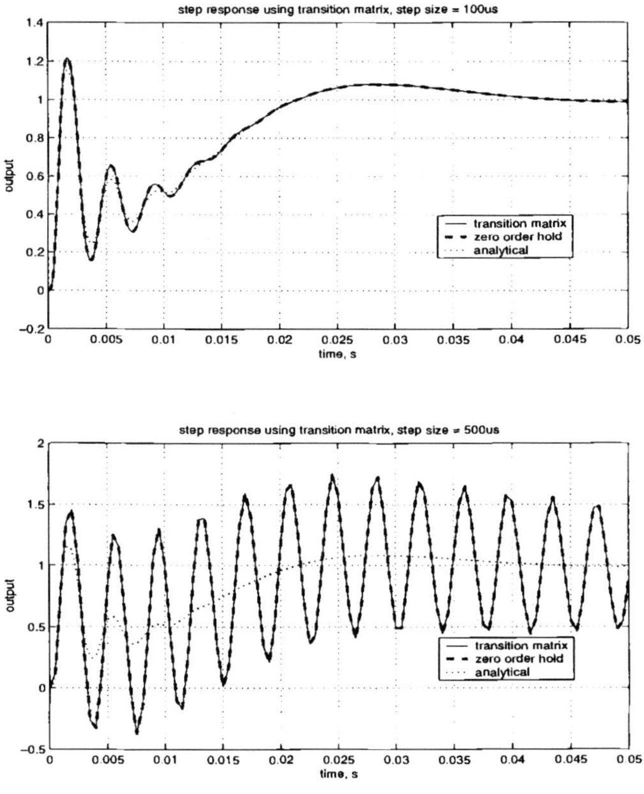

Figure 6 - *Step response of spring system with a gain of 10 for different step sizes.*

Implications for Digital Controllers

For a real system with an analog controller, the error is continually updated. Thus the dSpace simulation for larger step sizes is not correct and there may be differences between the simulation and test. For a digital controller, the dSpace simulation should be accurate if the integration time is less than the update rate of the controller, since the output is constant over the sampling period.

Since the dSpace simulation is less stable than the analytical (analog) solution, this implies that a digital controller may reduce the stability of a system. While this may not be expected for structural tests where specimen natural frequencies are typically less than 10 Hz, it can occur due to high natural frequencies of the specimen/actuator combination [2].

Structural Test Example

State Space Model of System

To evaluate further the dSpace system in a closed loop situation more representative of a structural test, the example of a cantilever beam driven by two actuators, one at the tip and one at the mid span, was used. Again, the first step was to generate the state space matrices for the system. The open loop model for the system [*3*] is shown in Figure 7. The beam was represented with four modes and the parameters chosen to give the same first natural frequency as for the earlier spring structure. The actuators and valves were the same as in the previous example.

To verify the model prior to transferring the matrices to the dSpace environment, the step responses were calculated using the open loop matrices and the Euler and transition matrix integration schemes to simulate the responses that would be expected in the real-time calculations.

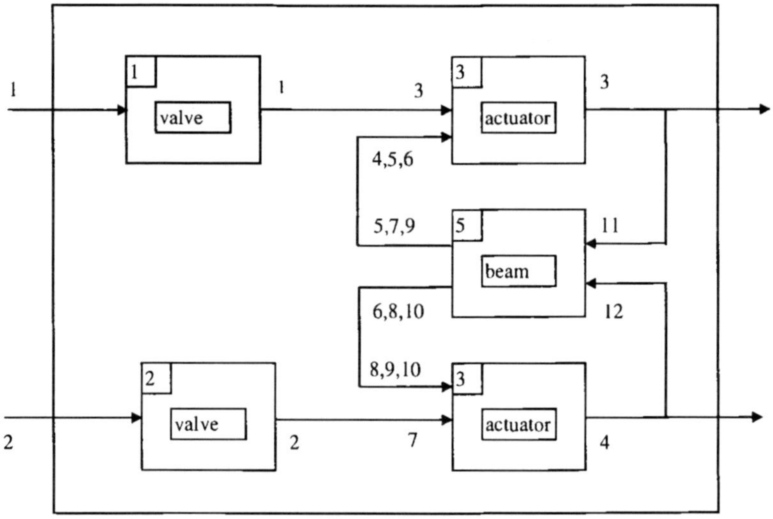

Figure 7 - *Open loop model of beam with two actuators.*

The responses at both the tip and mid span for a step input at the tip for an integration step size of 1 ms for the transition matrix method and 50 μs for the Euler solution are shown in Figure 8. The Euler solution is clearly unstable while the transition matrix method produces a response that is very similar to the analytical solution, which is shown as a dotted line in the figure. A real-time solution using transition matrix integration and a step size of 1 ms should therefore be stable. Step responses using integration step sizes of 100 μs and 250 μs for the transition matrix method are shown in Figure 9 for comparison. With a step size of 100 μs, the step response is almost indistinguishable from the exact solution and even at 250 μs the result is very close.

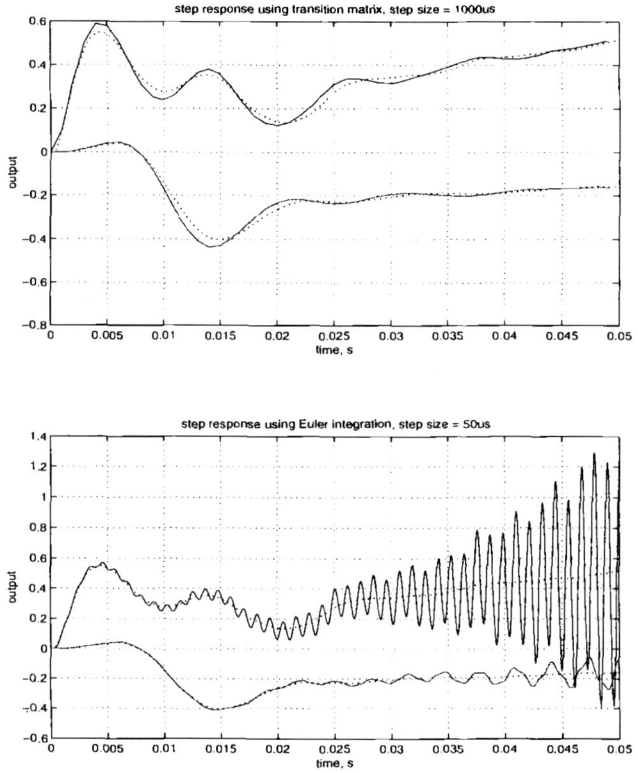

Figure 8 -*Response of beam system to step input at tip using transition matrix and Euler integration.*

Real-Time Implementation

The Φ matrix for this example is 16 by 16, while the *C* and Γ matrices are 2 by 16 and 16 by 2 respectively and were simply pasted into the dSpace program. It was then compiled, linked, and downloaded with the /o option to generate optimized code.

A function generator was connected to the program input of a servo controller. The error signal was then fed to input #1 on the real-time controller board. The error signal from a second servo controller with a zero command input was fed to input #2. Both outputs from the controller board were connected to an oscilloscope and the external conditioner inputs of the respective controller.

Using square and sine wave inputs to the tip actuator yielded the expected responses at both actuators. With a high gain on the mid span actuator, the mid span actuator load remained near zero as expected. With a low gain, the load on the mid span actuator was quite large and of opposite sign to the tip actuator load. Thus the system provides a good example of the interaction between actuators on a structure and again shows promise as an educational tool.

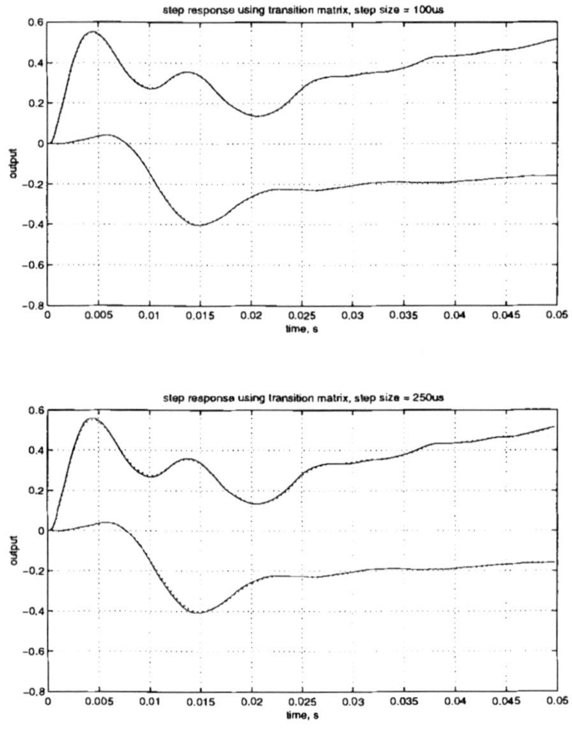

Figure 9 - *Response of beam system to step input at tip using transition matrix integration.*

Execution Times and Stability

The execution time for this two channel, four-mode system was less than 100 μs using optimized code. Instability occurred at a gain of about 1 using a 1 ms step size and at about 7 when using a step size of 100 μs.

Conclusions

It has been demonstrated that it is possible to simulate the response of a cantilever beam with a servo hydraulic actuator at the tip and mid span in real time in conjunction with an analog controller. Representing the beam with four modes results in a total of 16 states for the system. Execution times for this system are less than 100 μs while stable solutions can be achieved with integration step sizes of greater than 1 ms. It therefore appears feasible to simulate more complex systems with this simple processor. Very large systems will require faster processors working in parallel.

Large integration step sizes require the use of the transition matrix solution method, which makes use of the linearity of the system. This solution technique gives accurate solutions for any step size if the input is constant. However, for an analog system, the input is continuous and the simulation may be less stable than a real test. The input to the system using a digital servo controller is constant over the update period. The dSpace simulation should therefore reproduce a real test using a digital servo controller more accurately as long as the integration step size is less than or equal to the update rate of the controller.

At its present level of development, the system shows excellent potential for use as an educational tool. When interfaced with a servo controller, it is possible to demonstrate the effects of the various control parameters as well as the interaction effects of actuators. With additional development, it may be possible to provide the test engineer with a "virtual test" that can be tuned prior to setting up the physical test.

Acknowledgments

This work was funded by MTS Systems Corporation under a research contract with the National Research Council Canada.

References

[1] Hewitt, R. L. and Albright, F. J., "Computer Modeling and Simulation in a Full-Scale Aircraft Structural Test Laboratory," *Applications of Automation Technology to Fatigue and Fracture: Testing and Analysis: Third Volume, ASTM STP 1303*, A. A. Braun and L. N. Gilbertson, Eds., American Society for Testing and Materials, Philadelphia, 1997, pp. 81-95.

[2] Hewitt, R. L., "State Space Modeling and Experimental Verification of a Single Channel, Moving Load Cell Structural Test," *International Journal of Fatigue*, Vol. 22, No. 9, 2000, pp. 767-780.

[3] Hewitt, R. L., "State Space Modeling of a Multi-Channel, Moving Load Cell Structural Test," to be published.

[4] Meirovitch, L. *Elements of Vibration Analysis*, Second Edition, McGraw-Hill, Inc., New York, 1986, pp. 478-483.

Life Estimation

William T. Riddell[1]

On the Use of Numerical Models to Design Fatigue Crack Growth Tests for a Railroad Tank Car Spectrum

---

Reference: Riddell, W. T., **"On the Use of Numerical Models to Design Fatigue Crack Growth Tests for a Railroad Tank Car Spectrum,"** *Applications of Automation Technology in Fatigue and Fracture Testing and Analysis: Fourth Volume, ASTM STP 1411*, A. A. Braun, P. C. McKeighan, A. M., Nicholson, and R. D. Lohr, Eds., American Society for Testing and Materials, West Conshohocken, PA, 2002.

Abstract: The railroad tank car industry is moving toward damage-tolerance-based inspection intervals to maintain the safe and efficient transport of hazardous materials by rail. One aspect of this effort is predicting fatigue crack growth rates that result from railroad tank car service conditions. A fracture mechanics parameter, *H*, is proposed to relate the crack-tip driving force under a railroad tank car spectrum to the resulting crack growth rates. This approach will allow the direct application of fatigue crack growth rate data resulting from laboratory tests to models for crack growth under railroad service conditions. However, for this approach to be successful, it is essential that similitude between lab tests and the structural application is maintained, so that fatigue crack growth rates are predicted correctly. Two crack-closure-based models are used to predict the effects of various laboratory test design parameters on fatigue crack growth rates for the loaded vertical coupler force (LVCF) spectrum, which describes the vertical coupler forces for a loaded railroad tank car. Random ordered load blocks are predicted to result in growth rates approximately twice as great as those resulting from either low-high or high-low ordered blocks. Truncating even the smallest amplitudes from the spectrum was predicted to affect crack growth rates. The most significant effect predicted was that of the far-field stress level. Although changes in stress level are predicted to have only a minor affect (approximately 20%) on constant-amplitude fatigue crack growth rates, corresponding changes in stress level are predicted to affect variable-amplitude fatigue crack growth rates by up to a factor of six. These results suggest that tests designed to determine fatigue crack growth rates for the LVCF spectrum must be considered carefully, and that crack closure based models for fatigue crack growth are a valuable tool for designing such tests.

Keywords: variable amplitude, fatigue crack growth, railroad, tank car, fatigue crack closure, laboratory test design

---

[1]Mechanical Engineer, Structures and Dynamics Division, DTS-76, Volpe National Transportation Systems Center, 55 Broadway, Cambridge, MA 02142.

Much of the hazardous material that is transported in the United States is shipped in railroad tank cars. Although rail transport of hazardous materials typically results in only a few releases with relatively minor consequences [*1*] each year, the potential consequences of a severe accident are such that the Federal Railroad Administration (FRA) strives to continually improve the safety of hazardous materials transport by rail. One cause of hazardous materials release is fatigue failure. A number of fatigue failures occurred in the stub sill region of tank cars during the late 1980s and early 1990s [*2*]. As a result, periodic inspections of railroad tank cars were instituted in 1992 to identify cars with cracks so that these cars could be repaired or taken out of service before a fatigue failure occurred [*3*]. Although inspection and maintenance practices have reduced the occurrence of fatigue failures in recent years, there is a high cost involved with these practices, and there is no guarantee that the current inspection interval will continue to provide safe operation as the tank car fleet continues to age. Therefore, the tank car industry and the Federal Railroad Administration are developing damage tolerance methodology to establish performance-based inspection strategies for tank cars that will enhance the safety and efficiency of hazardous materials transport by rail.

A damage tolerance approach identifies a safe inspection interval for a fatigue-critical location by calculating the time it takes the largest crack that might be missed in an inspection to grow to a size that will cause a failure. Typically, a numerical model is used to predict fatigue crack growth rates. The FRA and the tank car industry are working to develop models for fatigue crack growth rates in railroad tank cars. For example, efforts to develop finite-element models for stress analyses, determine residual stresses resulting from manufacture, improve confidence in the description of load history, and develop databases of constant-amplitude fatigue crack growth characteristics for steel alloys used in the manufacture of tank cars are ongoing. The capability to predict fatigue crack growth under the variable-amplitude spectrum encountered by railroad tank cars is an additional technical hurdle that must be overcome. This paper will discuss efforts to develop a practical, laboratory-test-based methodology to predict variable-amplitude fatigue crack growth rates in railroad tank cars.

In theory, variable-amplitude fatigue crack growth rates could be predicted using a mechanism-based model for load interaction to simulate crack growth under the desired load spectrum. However, direct application of the desired load spectrum to laboratory coupon test specimens is thought to be more likely to result in reliable fatigue crack growth rate predictions for actual tank cars. Unlike constant-amplitude tests, which can be carefully governed using the ASTM Standard Method for the Measurement of Fatigue Crack Growth Rates (E647) there are no ASTM standards explicitly for the design of variable-amplitude fatigue crack growth rate tests. Therefore, it is important that variable-amplitude fatigue crack growth behavior is modeled prior to the design of laboratory tests so that effects that might compromise the similitude between laboratory coupons and the desired application are identified. The purpose of this paper is to utilize existing crack-closure-based models for fatigue crack growth rates under variable-amplitude loading to identify aspects of test design that can affect the resulting fatigue crack growth rates.

Crack Growth under Railroad Tank Car Loading Spectrum

The Association of American Railroads (AAR) has developed a series of load spectra to characterize longitudinal and vertical coupler forces, as shown in Figure 1, encountered under normal railroad tank car service conditions [*4*]. Each spectrum consists of a series of maximum and minimum load pairs, as well as the number of times that each pair is assumed to occur over 38 927 km of service. No information on load order or time history is given. In this paper, the loads encountered over 38 927 km of service are collectively called a "block." In most fatigue-critical locations on a tank car, the vertical coupler forces are the most damaging fatigue cycles. Therefore, the loaded vertical coupler force spectrum (LVCF) will be discussed in this paper. The set of loads defined as the LVCF spectrum is denoted by $\hat{F}$. The full spectrum has 1 964 178 load excursions. These excursions are sorted by amplitude into "bins" in 11.12 kN increments. Cumulative load exceedances for this spectrum are plotted in Figure 2. The cumulative exceedances are the number of cycles remaining after each successive bin has been truncated. The loads in the spectrum can be ordered, truncated and/or clipped to establish a modified block that will facilitate laboratory testing or numerical analysis. As an example, the spectrum was truncated to 33.36 kN and given a random order. A small portion of this truncated and random-ordered spectrum is plotted in Figure 3.

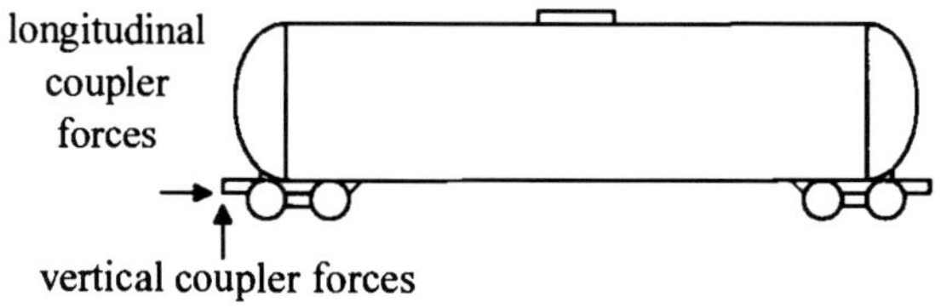

Figure 1. *Load cases for damage tolerant analysis of a railroad tank car.*

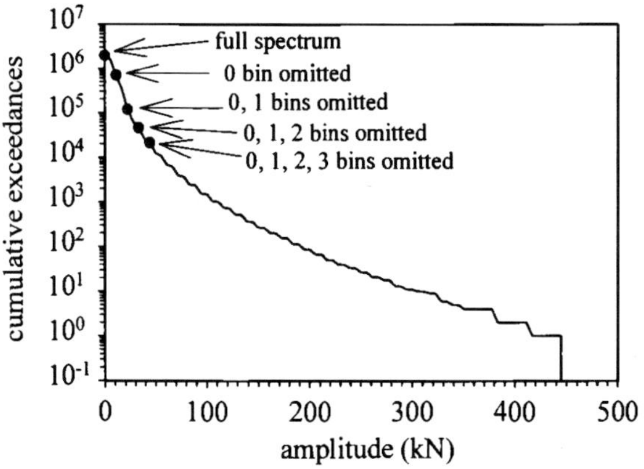

Figure 2. *Cumulative load exceedances for vertical coupler forces.*

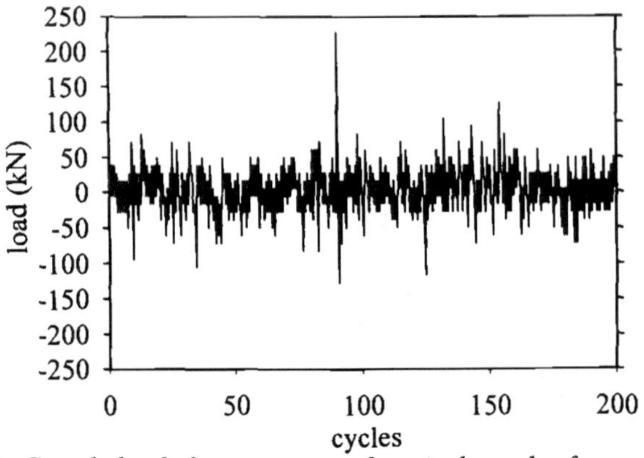

Figure 3. *Sample loads from a truncated vertical coupler force spectrum.*

Often, a spectrum must be truncated or clipped to allow practical laboratory testing or numerical analysis. A modified block of loads is intended to be an approximation of the complete spectrum. However, three of the most common methods to describe crack tip driving force under variable-amplitude loading conditions: maximum K, average ΔK and RMS average ΔK, change when the spectrum is truncated or clipped. Therefore, a new parameter to describe the crack tip driving force under a variable-amplitude load spectrum is developed. The value of this parameter does not change if the spectrum is truncated or clipped. Therefore, this parameter can be used to relate crack growth rates resulting from a modified block of loads to crack growth rates for the full spectrum.

From the time the Paris model [*5*] for fatigue crack growth rates was suggested, nearly all models for fatigue crack growth have utilized the stress intensity factor as the crack-tip driving force. Predicting fatigue crack growth rates that correspond to a series of far-field load cycles requires that these loads be converted to a series of crack-tip stress intensity factor cycles. To achieve this, a global structural analysis is used to relate a local stress component, σ, to a far-field applied load, F. This relation can be described by

$$\sigma = \Omega F \tag{1}$$

where Ω is the stress scale factor, which depends on the configuration of the tank car. A fracture mechanics analysis relates the stress intensity factor to the local stress component. Typically, the relationship takes the form

$$K = \sigma \beta(a) \sqrt{\pi a} \tag{2}$$

where a is the crack length and $\beta(a)$ is a function of geometry, crack size and crack shape [*6*]. Combining Equations 1 and 2 results in the following relationship.

$$K = \Omega F \beta(a) \sqrt{\pi a} \tag{3}$$

Inspecting Equation (3) suggests that the term

$$H = \Omega\beta(a)\sqrt{\pi a} \tag{4}$$

can relate a series of far-field load excursions, $\hat{F}$, to a series of crack-tip stress intensity factor excursions, $\hat{K}$, by

$$\hat{K} = H\hat{F} \tag{5}$$

In this paper, crack growth rates are described by crack growth per block (da/dB). To the extent that a series of stress intensity factor cycles, described by $\hat{K}$, will govern fatigue crack growth rates, the parameter H must also govern fatigue crack growth rates for a given load spectrum, $\hat{F}$. Furthermore, the value of H does not change if the full spectrum is approximated by a truncated or clipped spectrum, making it an ideal description of primary crack-tip driving force for this application. Therefore, excepting secondary effects, crack growth rates for the LVCF spectrum should result in a single curve when plotted against H.

Results from a series of fatigue crack growth rate tests are considered [7]. In these tests, load blocks representing the LVCF spectrum were applied repeatedly to edge-cracked coupon test specimens, and the resulting crack lengths were recorded. The tests had various levels of stress scale factor, Ω (low = 7.75×10^{-4} mm^{-2}, high = 1.04×10^{-3} mm^{-2}) and truncation (low = 0 bin omitted, medium = 0,1 bins omitted, high = 0, 1, 2 bins omitted) applied. The load spectrum for one test was clipped to limit the maximum applied loads. A curve of the form

$$blocks = c_0 + \frac{c_1}{a} + \frac{c_2}{a^2} + \frac{c_3}{a^3} + \frac{c_4}{a^4} + \frac{c_5}{a^5} \tag{6}$$

was fitted to crack length data from each test, where a is crack length, blocks represents applied loads, and c_i are the curve fit coefficients. Equation (6) was differentiated to determine crack growth rate per block (da/dB) as a function of a. The crack growth rate per block is plotted against a in Figure 4. Equation (3) was used to convert these results to a da/dB-H relationship. The resulting data are plotted in Figure 5. Although there is some scatter in the da/dB-H data, the general trend is a single curve, which appears analogous to classic constant-amplitude behavior with a linear Paris regime ($H > 0.005$ mm$^{-1.5}$) and a near-threshold regime ($H < 0.005$ mm$^{-1.5}$). Spikes in some of these data, denoted by A on the graph, result from load control errors that occurred during the tests. The transient effect that is especially prominent at the start of one test, denoted by B on the graph, will be discussed later in the paper.

In practice, the relationship between crack size and H can be calculated using finite element, influence function, or table-look-up techniques by utilizing the similarity of the form of H to the form of K: inputting the local stress component that results from a unit load to a code such as NASGRO [8] will result in the correct value of H (but will return units of force length$^{-1.5}$ rather than length$^{-1.5}$). The H-da/dB and a-H relationships can be combined to derive the a-da/dB relationship, which can be integrated numerically to establish a crack size against blocks (and therefore time) relationship. In this manner, engineers can utilize variable-amplitude fatigue crack growth rate data to make life predictions under the complex load spectrum encountered by railroad tank cars without performing the cycle-by-cycle analyses typically required to predict variable-amplitude fatigue crack growth rates. The challenge is to design laboratory tests that will result in fatigue crack growth rates that are representative of those that will occur under service conditions. The remainder of this paper will discuss two numerical modeling techniques that are used to study the effects of some of the laboratory test conditions on resulting variable-amplitude fatigue crack growth rates, and the results of these studies.

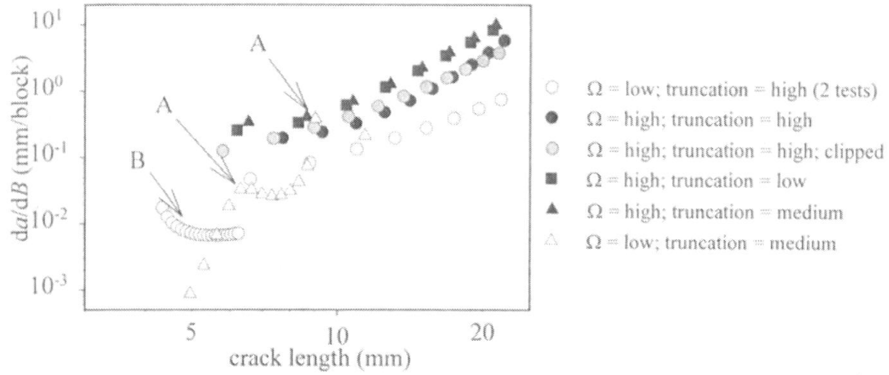

Figure 4. *Experimentally observed variable-amplitude fatigue crack growth rates for the LVCF spectrum in A572-Grade 50 steel plotted against crack length [7]. Some of the unusual-looking date in this plot are attributed to (A) load control errors and (B) transient effects at the start of a test.*

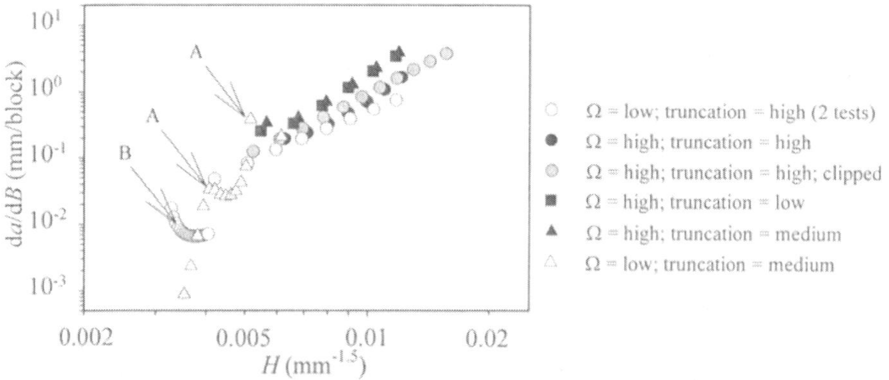

Figure 5. *Crack growth rates from Figure 4 plotted against fracture mechanics parameter, H. Some of the unusual-looking date in this plot are attributed to (A) load control errors and (B) transient effects at the start of a test.*

Models for Fatigue Crack Growth Rates

Crack closure, first proposed by Elber [*9, 10*], was a major breakthrough in the mechanistic modeling of many secondary effects on crack growth rate. Elber realized that crack faces near the crack tip can remain closed under nominally tensile loading conditions, and suggested that no fatigue damage occurs when the crack faces are in contact. This model retains the basic Paris model [*5*] dependency on ΔK by considering the effective cyclic stress intensity factor that is encountered by the crack tip, which is defined by

$$\Delta K_{eff} = K_{max} - K_{open} \tag{7}$$

where K_{open} is the stress intensity factor at which the crack faces are first completely open. Sometimes the effective cyclic stress intensity factor is also described by

$$\Delta K_{eff} = U\Delta K \tag{8}$$

where

$$U = \frac{1 - \frac{K_{open}}{K_{max}}}{1 - R} \tag{9}$$

and R is the stress ratio. This form is useful when discussing closure in terms of the effect on fatigue crack growth rates.

In many circumstances, effective cyclic stress intensity factor allows fatigue data for different stress ratios to fall onto a single curve, whereas the same data will exhibit considerable scatter when plotted as da/dN against ΔK [*11*]. The resulting fatigue crack growth rate relationship is often described by

$$\frac{da}{dN} = c\left(\Delta K_{eff}\right)^n \tag{10}$$

in the Paris regime, where c and n are not necessarily the same values as for the Paris model. The unique relationship between ΔK_{eff} and da/dN suggests that, under many conditions, ΔK_{eff} is a good description of the mechanical crack tip driving force. Plasticity [*9, 10, 12*], roughness [*13-15*], and oxide layers or debris [*15-17*] have all been identified as potential sources of fatigue crack closure. Opening loads have been studied using theoretical [*18*], numerical [*19, 20*], experimental [*21*] and hybrid [*22*] approaches. However, there is still some debate regarding the cause, measurement, prediction and implication of fatigue crack closure [*23*].

Fatigue crack closure has also been proposed as the controlling mechanism for crack growth acceleration and retardation following over- and under-loads in variable-amplitude loading conditions [*24, 25*]. The ΔK_{eff} model for crack growth suggests that fatigue crack growth rates under variable-amplitude loading conditions can be predicted from da/dN-ΔK_{eff} data developed under constant-amplitude loading conditions, if the opening loads throughout the load history are known. Variable-amplitude fatigue crack growth rates in this paper are predicted using a simple spreadsheet model and the computer code FASTRAN [*26*]. Both models employ ΔK_{eff} as the primary crack-tip driving force. For the analyses discussed in this paper, fatigue crack growth rates for individual cycles are predicted from ΔK_{eff} values by Equation (10), with no fatigue crack growth assumed to occur if ΔK_{eff} is less than ΔK_{th}. The parameters used for Equation (10) in this paper are summarized in Table 1. These values are reasonable for steel [*7*, *27*].

Table 1. -- *Parameters for fatigue crack growth analyses.*

c	2.74×10^{-9}
n	3.32
ΔK_{th}	0.00

Average Opening Load Model

The average opening load model (AOLM) is a spreadsheet-based algorithm that is used to model the effect of truncation on fatigue crack growth rates under the LVCF spectrum. Similar models have been used elsewhere [*28*]. Rather than calculating the opening load throughout the load history, a single opening load ratio, K_{open}/K_{max}, is assumed to characterize the entire block. An average opening load ratio of 0.00 is used for the analyses discussed herein, although different values of this ratio could be used. Despite a somewhat crude treatment of crack closure, this model can give insight into the sensitivity of da/dB to the effects of load truncation.

FASTRAN Analyses

FASTRAN uses Newman's modified Dugdale model to calculate opening loads [*30*]. The modified Dugdale model, also known as a strip-yield model, accounts for plasticity-induced closure (PICC), but neglects other potential load interaction effects. Despite considering only plasticity-induced crack closure, FASTRAN has been used to successfully predict opening loads to reduce fatigue crack growth rates resulting from constant-amplitude tests with different stress ratios to a single da/dN-ΔK_{eff} curve [*30*] and to predict variable-amplitude fatigue crack growth rates from such curves [*24-25*].

A typical plot of crack length versus blocks (a-B) resulting from a FASTRAN analysis of a LVCF spectrum is shown in Figure 6. The coefficients for Equation (6) were fitted to the prediction, and differentiated to find da/dB. The calculated crack growth rates (da/dB) are plotted against H in Figure 7. The code predicts an initial period where da/dB decreases as a increases. These data (open symbols in Figure 5b) result from a delayed buildup of crack closure, and are a transient effect, which can also be observed in laboratory data (see open circles for $0.003 \text{ mm}^{-1.5} < H < 0.004 \text{ mm}^{-1.5}$ in Figure 5). The remaining data (closed symbols) result in a nearly linear relationship between da/dB and H, which agrees with the experimentally observed trend plotted in Figure 4 for H in the analogous Paris regime.

FASTRAN requires the input of the α parameter, which describes the crack-tip constraint. A value of 1.0 represents plane stress conditions, while a value of 3.0 represents plane strain conditions. The constraint parameter affects predicted opening loads, and therefore predicted fatigue crack growth rates. Typically, series of constant-amplitude laboratory tests are performed at various stress ratio (R) levels, and the α value that best collapses the data to a single curve is then used for numerical variable-amplitude crack growth rate models. Lower values of ΔK tend to correlate best with plane strain α values (in the range of 2.0 to 2.5). Higher values of ΔK tend to correlate best with plane stress α values (near 1.0). Newman has had good success correlating fatigue crack growth rates for different stress ratios over a wide range of ΔK by linking a transition from a plane strain to a plane stress α value to a change from flat to slant fatigue crack growth conditions [*31*].

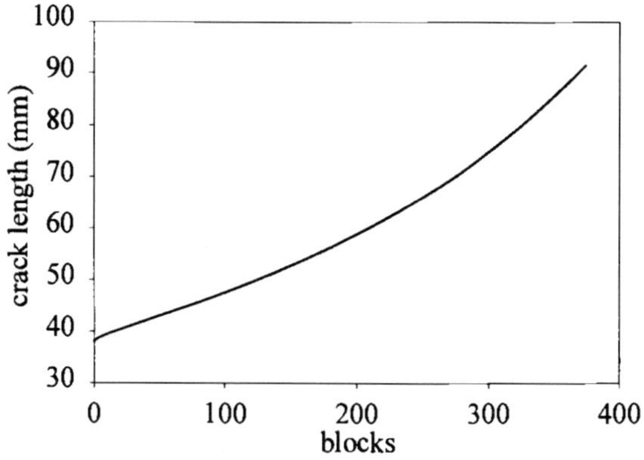

Figure 6. *Results of variable-amplitude fatigue crack growth simulations. Predicted crack length is plotted agaist applied blocks.*

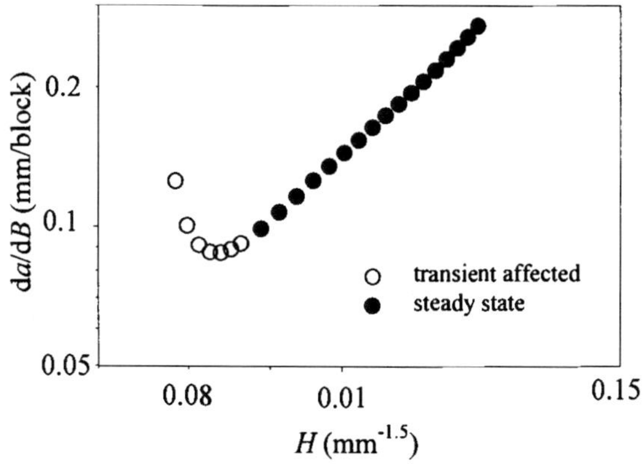

Figure 7. *Results of variable-amplitude fatigue crack growth simulations. Predicted crack growth rate is plotted against variable-amplitude driving force, H.*

To study the effect of α on variable-amplitude fatigue crack growth rates, a series of analyses of the LVCF spectrum (similar to those for which the results are shown in Figure 5) were performed for identical geometric and loading configurations, but different α values. The fatigue crack growth rate predicted at $H = 0.011$ mm$^{-1.5}$ for various values of α was scaled by the crack growth rate predicted at $H = 0.011$ mm$^{-1.5}$ for $\alpha = 3.0$. These results are plotted in Figure 8 (solid line). For comparison, the effect of α on constant-amplitude fatigue crack growth rates was considered. For constant-amplitude loading conditions, the effect of α on predicted fatigue crack growth rates can be studied by substituting published equations for opening load as a function of α and σ/σ_y [8] into Equations (8) and (10), resulting in crack growth rates described by

$$\frac{da}{dN} = c(U\Delta K)^n \tag{11}$$

in the Paris regime. Then, the ratio of crack growth rates for a given α, $(da/dN)_\alpha$, to the rate for $\alpha = 3.0$, $(da/dN)_{\alpha=3.0}$, is predicted to be

$$\frac{\left(\frac{da}{dN}\right)_\alpha}{\left(\frac{da}{dN}\right)_{\alpha=3.0}} = \left(\frac{U_\alpha}{U_{\alpha=3.0}}\right)^n \tag{12}$$

in the Paris regime, where U_α is the value for a given value of α and $U_{\alpha=3.0}$ is the value for $\alpha = 3.0$. The results of Equation (12) are also plotted in Figure 8 (dashed line). These results suggest that α has a similar effect on predicted fatigue crack growth rates for both constant- and variable-amplitude fatigue crack growth rates.

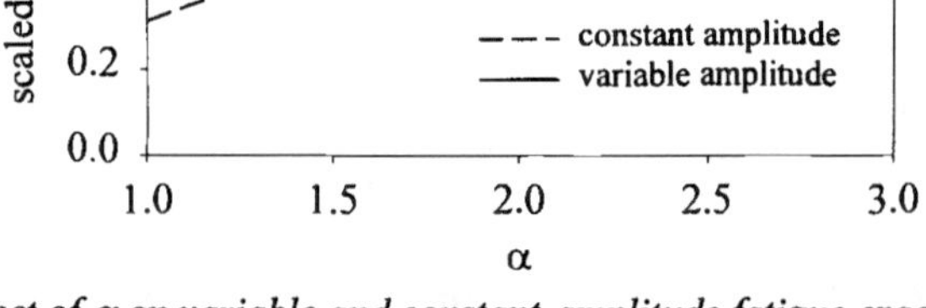

Figure 8. *Effect of α on variable and constant-amplitude fatigue crack growth rates.*

Results and Discussion

Truncation

The full LVCF spectrum has 1 964 178 load cycles. Therefore, practical laboratory testing or mechanistic numerical analysis is likely to employ truncation to some degree. The fatigue crack growth rate predicted for blocks at a given values of *H* but with different various levels of truncation were compared. The predicted effect of truncation on fatigue crack growth rates at $H = 0.00624\ mm^{-1.5}$ is shown in Figure 9. These results are consistent with values for values of *H* that cause fatigue crack growth rates of practical interest to damage tolerance (greater than 0.1 mm/block, which for many tank cars will translate to approximately 0.1 mm/year). The results of these analyses suggest that any degree of truncation to the experimentally applied spectrum could affect the resulting fatigue crack growth rates. Although it is possible that using a higher opening level and threshold value will cause the true effects of truncation to be less than that shown in the figure, there is not an obvious level of truncation for which it will be safe to assume no effect on fatigue crack growth rates. Therefore, efforts will have to be made to test for truncation effects or to bound these effects using numerical models and apply an appropriate safety factor.

One approach to designing laboratory tests to determine fatigue crack growth rates for the LVCF spectrum is to utilize different truncation levels for different *H* regimes. On one extreme, the LVCF threshold crack growth rates should be dependent only on the maximum and minimum load encountered during the full block. Therefore, near-threshold spectrum fatigue crack growth behavior should approach constant-amplitude threshold behavior for the appropriate stress ratio. On the other extreme, the crack growth rates resulting for large values of *H* are such that significant crack growth will occur in less than a full spectrum. Here, an abbreviated spectrum that contains the correct ratio of small amplitude load cycles could be applied. A difficult question in this case is how to treat the large amplitudes that occur only relatively infrequently during the block. Overlap of different truncation levels for the same value of *H* can be used to bound the effects of truncation on fatigue crack growth rate.

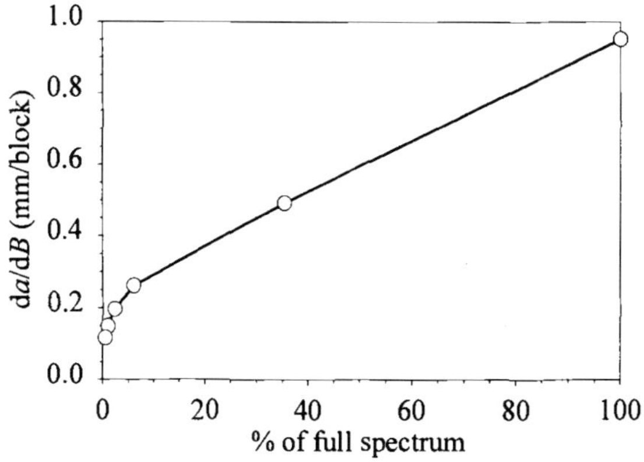

Figure 9. *Predicted effect of truncation on crack growth rates for* $H = 0.00624\ mm^{-1.5}$.

Load Order

The effect of load order on fatigue crack growth rates was studied. The LVCF spectrum was truncated such that there were 44 948 cycles per block. Three different random orders were generated, and the spectrum was also ordered from lowest to highest and then from highest to lowest load. In a random order, two consecutive cycles are sometimes combined into a single larger cycle. Therefore, the random blocks actually contain between 42 951 and 43 001 cycles. Each resulting block was applied to a virtual test specimen repeatedly. The effects of load order on fatigue crack growth rates are thought to be largely a function of varying crack opening levels throughout the load history. Therefore, FASTRAN is an appropriate tool to study load order effects on fatigue crack growth rates. An α value of 3.0 was used for these analyses. Although the constraint parameter α and the crack growth model coefficients c, n and ΔK_{eff} affect the absolute fatigue crack growth rates, the trends discussed herein are relatively insensitive to reasonable choices of these parameters.

FASTRAN-predicted crack growth rates resulting from the five different load histories are plotted against H in Figure 10. As the effects of load order were of interest in these analyses, the fatigue crack growth rates shown in Figure 10 were not obtained by taking the derivative of an equation of the form shown in (6) fitted to data. Rather, rates were obtained by considering $\Delta a/\Delta B$ directly, where ΔB values were less than a single block for the larger crack growth rates. This approach was taken to highlight any potential variations in fatigue crack growth rate that might occur within blocks that result from ordering the load cycles. However, this approach results in the steps in crack groth rates that are especially evident for both the low to high and high to low ordered blocks for H greater than 0.006 mm$^{-1.5}$. All three random orders result in crack growth rates that are faster than both the high-low and low-high ordered blocks. Although this effect (a factor of two in fatigue crack growth rates) might be small enough so as to fall within experimentally observed scatter, both the low-high and high-low orders result in crack growth rates that are nonconservative compared to the rates for the random load orders. Therefore, the loads in the spectrum should be ordered randomly to avoid improper and nonconservative results.

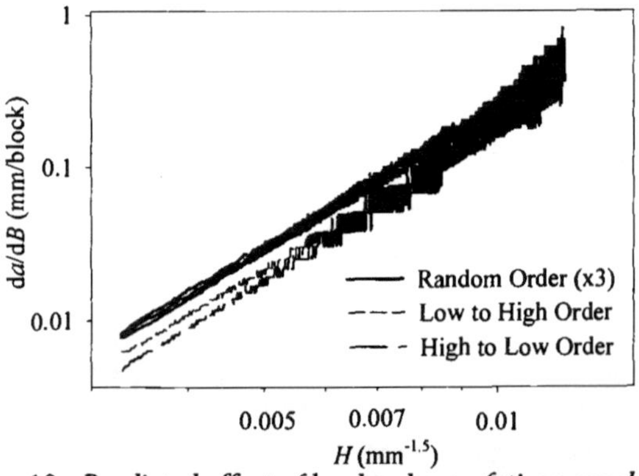

Figure 10. *Predicted effect of load order on fatigue crack growth.*

Stress Scale Factor

The parameter, *H*, coupled with the load history, $\hat{F}$, defines the stress intensity factor history, $\hat{K}$. However, there are different combinations of configuration, crack length, and stress scale factor that can result in the same value of *H*. Although stress intensity factor history alone is often considered to determine the crack growth rate, the equations for crack opening load that were put forth by Newman for constant-amplitude fatigue crack growth rates [*28*] include an effect of far-field stress levels.

To consider the effects of far-field stress level (defined by the stress scale factor Ω) on variable-amplitude fatigue crack growth rates, a series of FASTRAN analyses were performed for the LVCF spectrum with values of Ω ranging from approximately 0.15 to 0.75 times Ω_y, where Ω_y is defined as the value for which far-field yielding would occur under the highest tensile load in the load spectrum. The predicted fatigue crack growth rate at $H = 0.011$ mm$^{-1.5}$ was recorded for each analysis. The resulting predicted crack growth rates are scaled by the rate predicted for $\Omega/\Omega_y = 0.15$, and plotted against Ω/Ω_y in Figure 11. Also plotted are the scaled fatigue crack growth rates predicted by Newman's equation for constant-amplitude crack growth ($\alpha = 3.0$) and a Paris exponent of 3.32. The far-field stress level is predicted to have only a minor effect (approximately 20%) on constant-amplitude fatigue crack growth rates in the range considered. As a result of the predicted low sensitivity to the ratio of maximum stress to yield stress for constant-amplitude fatigue crack growth, this ratio is set to 1/3 for most materials in the NASGRO material property database and not changed. However, the far-field stress level is predicted to effect variable-amplitude fatigue crack growth rates by a factor of six in the range of Ω investigated. In other words, not only the stress intensity factor history (described by *H* for the LVCF spectrum), but also the far-field load level has a strong effect on the variable-amplitude fatigue crack growth rates for the LVCF spectrum. These results suggest that the concept of similitude, where the stress intensity factor history governs fatigue crack growth rates, is much less robust for variable-amplitude loading than for constant-amplitude loading. Therefore, the stress scale factor in laboratory tests must be similar to that expected under service conditions.

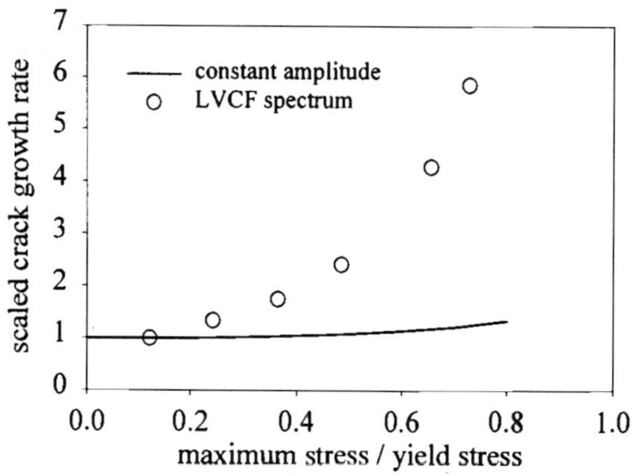

Figure 11. *Predicted effect of stress scale factor on crack growth rates.*

Summary and Conclusions

The direct application of a desired load spectrum to laboratory test coupons results in fatigue crack growth rates that can be plotted as crack growth per block (da/dB) against H, a fracture-mechanics-based parameter that describes the primary crack-tip driving force for a given load spectrum. Once established, the resulting relation can allow engineers to perform damage tolerance analyses for variable-amplitude loading conditions without performing cycle-by-cycle analyses typically required for variable-amplitude fatigue crack growth predictions. This approach can allow inspection intervals to be established for the many different fatigue critical locations on a railroad tank car with relatively little engineering effort on the part of car owners or operators. However, this approach requires that similitude between the laboratory coupons and the tank car structure is maintained to ensure that the crack growth rates measured in the laboratory will reflect those in a tank car.

Numerical fatigue crack growth rate analyses of a railroad tank car spectrum suggest that variable-amplitude growth rates can be sensitive to laboratory test design. In some regards, crack growth rates from variable-amplitude load tests are more sensitive to test design than crack growth rates from constant-amplitude load tests are. Load order, truncation level, and far-field stress level are all predicted to affect the resulting fatigue crack growth rate. Random ordered load blocks are predicted to result in growth rates approximately twice as great as those resulting from either low-high or high-low ordered blocks. For the LVCF spectrum, it is predicted that any level of truncation will result in a change in crack growth rate. The most significant effect predicted was that of the far-field stress level. Although stress level is predicted to have only a minor effect (approximately 20%) on constant-amplitude fatigue crack growth rate, stress level is predicted to affect variable-amplitude fatigue crack growth rates by up to a factor of six for a comparable range in far-field stress levels. Taken as a whole, these results suggest that tests designed to determine fatigue crack growth rates for the LVCF spectrum must be considered carefully to ensure that resulting fatigue crack growth rates are representative of those that will be encountered in service. Crack-closure-based models for fatigue crack growth rate prediction are valuable tools for designing such tests.

Acknowledgments

The author thanks Dr. P.C. McKeighan for supplying the raw data used to develop Figures 4 and 5, and for useful comments and discussions in developing the numerical models and overall approach. This research was supported by the Federal Railroad Administration (FRA), Office of Research and Development. Ms. Claire Orth is the FRA program manager for Equipment Safety, and Mr. Jose Pena administers FRA research programs related to tank car safety.

References

[*1*] "Hazardous Materials Transportation in Tank Cars – Analysis of Risks – Part 1," *DOT/FRA/ORD-92/34*, Federal Railroad Administration, Washington, DC, 1992.

[2] "Inspection and Testing of Railroad Tank Cars," *NTSB/SIR-92/05*, National Transportation Safety Board, Washington, DC, 1992.

[3] Stone, D. H., Cardinal, J. W., "Application of Damage Tolerance Principles to Stub Sill Tank Car Cracking," *Proceedings of the 6th International Heavy Haul Conference, Cape Town, South Africa, 1997*, International Heavy Haul Association, Virginia Beach, VA, 1997, pp. 726-733.

[4] Cogburn, L. T., "Stub-Sill Tank Car Research Project – Results of a 15,000 mile Over the Road Test," *FRA/ORD/95-11*, Federal Railroad Association, Washington, DC, 1995.

[5] Paris, P. C., Gomez, M. P., Anderson, W.E., "A Rational Analytic Theory of Fatigue," *The Trend in Engineering*, 1961.

[6] Broek, D. "Elementary Fracture Mechanics," Fourth Revised Edition, Kluwer Academic Publishers, Dordrecht, 1986.

[7] McKeighan, P. C., Smith, K. B., Cardinal, J. W., "Load Interaction Effects in Coupon Fatigue Tests Using a Railroad Tank Car Spectrum," *Rail Transportation 1997, ASME: RTD-Vol 13,* American Society of Mechanical Engineers, New York, NY, 1997, pp. 33-41.

[8] Forman, R. G., Shivakumar, V., Mettu, S. R., Newman, J. C., Jr., "Fatigue Crack Growth Computer Program NASGRO Version 3.0 User's Manual," *JSC-22267B*, NASA Johnson Space Flight Center, Houston, TX, 1996.

[9] Elber, W., "Fatigue Crack Closure Under Cyclic Tension," *Engineering Fracture Mechanics,* Vol. 2, 1970, pp.37-45.

[10] Elber, W., "The Significance of Fatigue Crack Closure," *Damage Tolerance in Aircraft Structures, ASTM STP 486*, American Society of Testing and Materials, Philadelphia, PA, 1971, pp. 230-242.

[11] Riddell, W. T., Piascik, R. S., "Stress Ratio Effects in AA 2024," *Twenty Ninth National Symposium on Fatigue and Fracture Mechanics, ASTM STP 1332*, T. L. Panontin and S. D. Sheppard, Eds., American Society for Testing and Materials, West Conshohocken, PA, 1999.

[12] Elber, W., "Crack Closure and Crack Growth Measurements in Surface Flawed Titanium Alloy Ti-6Al-4V," *NASA TN D-8010*, National Aeronautics and Space Administration, Washington, DC, 1975.

[13] Adams, N. J., "Fatigue Crack Closure at Positive Stresses," *Engineering Fracture Mechanics*, Vol. 4, 1972, pp. 543-554.

[*14*] Purushothaman, S., Tien, J. K., "A Fatigue Crack Growth Mechanism for Ductile Materials," *Scripta Metallurgica*, Vol. 9, 1975, pp. 923-926.

[*15*] Walker, N. and Beevers, C. J., "A Fatigue Crack Closure Mechanism in Titanium," *Fatigue of Engineering Materials and Structures*, Vol. 1, 1979, pp. 135-148.

[*16*] Endo, K., Komai, K., Matasuda, Y., "Mechanical Effects of Corrosion Products in Corrosion Fatigue Crack Growth of a Steel," *Bulletin of JSME*, Vol. 24, 1981, pp. 1319-1325.

[*17*] Suresh, S., Zamininski, G. F., Ritchie, R. O., "Oxide Induced Crack Closure: An Explanation for Near-Threshold Corrosion Fatigue Crack Growth Behavior," *Metallurgical Transactions*, Vol. 12A, 1981, pp. 1435-1443.

[*18*] Budiansky, B. and Hutchinson, J. W., "Analysis of Closure in Fatigue Crack Growth," *Journal of Applied Mechanics*, Vol. 45, 1978, pp. 267-276.

[*19*] R. C. McClung, "Finite Element Analysis of Specimen Geometry Effects on Fatigue Crack Closure," *Fatigue and Fracture of Engineering Materials and Structures*, Vol 17, No. 8, 1994, pp. 861-872.

[*20*] Fleck, N. A., Newman, J. C., Jr., "Analysis of Crack Closure Under Plane Strain Conditions," *Mechanics of Fatigue Crack Closure, ASTM STP 982*, J. C. Newman, Jr. and W. Elber, Eds., American Society for Testing and Materials, Philadelphia, PA, 1988, pp. 319-341.

[*21*] Donald, J. K., "A Procedure for Standardizing Crack Closure Levels," *Mechanics of Fatigue Crack Closure," ASTM STP 982*, J. C. Newman, Jr. and W. Elber, Eds., American Society for Testing and Materials, Philadelphia, PA, 1988, pp. 222-229.

[*22*] Dawicke, D. S., Shivakumar, K. N., Newman, J. C., Jr. "An Inverse Method for the Calculation of Through-Thickness Fatigue Crack Closure," *Fracture Mechanics: Twenty Second Symposium (Vol. II), ASTM STP 1131*, S. N. Atluri, J. C. Newman, Jr., I. S. Raju and J. S. Epstein, Eds., American Society for Testing and Materials, Philadelphia, PA, 1992, pp. 46-57.

[*23*] *Advances in Fatigue Crack Closure Measurement and Analysis, Second Volume ASTM STP 1343*, R. C. McClung and J. C. Newman, Eds., American Society for Testing and Materials, West Conshocken, PA, 1999.

[*24*] Hsu, C., Chan, K. K., Yu, J., "Effects of Thickness on Plasticity-Induced Fatigue Crack Closure: Analysis and Experiment," *Advances in Fatigue Crack Closure Measurement and Analysis: Second Volume, ASTM STP 1343*, R. C. McClung and J. C. Newman, Jr., Eds., American Society for Testing and Materials, West Conshohocken, PA, 1999, pp. 285-303.

[*25*] Newman, J. C. Jr., "Crack Growth under Variable Amplitude and Spectrum Loading in 2024-T3 Aluminum Alloys," *TMS-ASM Fall Meeting, High Cycle Fatigue of Structural Materials – The Professor Paul C. Paris Symposium*, Indianapolis, IN, September 16-18, 1997.

[*26*] Newman, J. C., Jr. "FASTRAN-II – A Fatigue Crack Growth Structural Analysis Program," *NASA TM 104159*, 1992.

[27] Finger, R. W., "Prediction Model for Fatigue crack Growth in Windmill Structures," *Effect of Load Spectrum Variables on Fatigue Crack Initiation and Propagation, ASTM STP 714*, D. F. Bryan and J. M. Potter, Eds. American Society for Testing and Materials, Philadelphia, PA, 1980, pp. 185-204.

[*28*] Socie, D. F., "Prediction of Fatigue Crack Growth in Notched Members Under Variable Amplitude Loading Histories," *Engineering Fracture Mechanics*, Vol. 9, pp. 849-865, 1977.

[*29*] Sunder, R., Seetharan, S. A., Bhaskaran, T. A., "Cycle Counting for Fatigue Crack Growth Analysis," *International Journal of Fatigue*, Vol. 6, pp. 147-156, 1984.

[*30*] Newman, J. C., Jr., "A Crack Opening Stress Equation for Fatigue Crack Growth," *International Journal of Fracture*, Vol. 24, No. 3, 1984, pp. R131-135.

[*31*] Newman, J. C., Jr., "Effects of Constraint of Crack Growth Under Aircraft Spectrum Loading," *Proceedigns of the Specialists' Conference dedicated to the 65th Birthday of Jap Schijve*, A. Beukers, Th. deJong, J. Sinke, A. Vlot, L. B. Vogelesang, Eds., Delft University Press, Delft, 1992.

Antonio C. O. Miranda,[1] Marco A. Meggiolaro,[1] Jaime T. P. Castro,[1]
Luiz F. Martha,[1] and Tulio N. Bittencourt[2]

Fatigue Crack Propagation under Complex Loading in Arbitrary 2D Geometries

---

Reference: Miranda, A. C. O., Meggiolaro, M. A., Castro, J. T. P., Martha, L. F., and Bittencourt, T. N., "**Fatigue Crack Propagation under Complex Loading in Arbitrary 2D Geometries**," *Applications of Automation Technology in Fatigue and Fracture Testing and Analysis: Fourth Volume, ASTM STP 1411*, A. A. Braun, P. C. McKeighan, A. M. Nicolson, and R. D. Lohr, Eds., American Society for Testing and Materials, West Conshohocken, PA, 2002.

Abstract: A reliable and cost effective two-phase methodology is proposed and implemented in two pieces of software to predict fatigue crack propagation in generic two-dimensional structural components under complex loading. First, the fatigue crack path and its stress intensity factor are calculated in a specialized finite-element software, using small crack increments. At each crack propagation step, the mesh is automatically redefined based on a self-adaptive strategy that takes into account the estimation of the previous step stress analysis numerical errors. Numerical methods are used to calculate the crack propagation path, based on the computation of the crack incremental direction, and the stress-intensity factors K_I, from the finite element response. An application example presents a comparison between numerical simulation results and those measured in physical experiments. Then, an analytical expression is adjusted to the calculated $K_I(a)$ values, where a is the length along the crack path. This $K_I(a)$ expression is used as an input to a powerful general purpose fatigue design software based in the local approach, developed to predict both initiation and propagation fatigue lives under complex loading by all classical design methods, including the *S-N*, the ε-*N* and the *IIW* (for welded structures) to deal with crack initiation, and the *da/dN* to treat propagation problems. In particular, its crack propagation module accepts any K_I expression and any *da/dN* rule, using a ΔK_{rms} or a cycle-by-cycle propagation method to deal with one and two-dimensional crack propagation under complex loading. If requested, this latter method may include overload-induced crack retardation effects.

Keywords: fatigue crack propagation, finite elements, arbitrary loading

---

[1] Ph.D. Student – Dept. of Civil Engineering, Research Scientist – Dept. of Mechanical Engineering, Associate Professor – Dept. of Mechanical Engineering, and Associate Professor – Dept. of Civil Engineering, respectively, Pontifical Catholic University of Rio de Janeiro (PUC-Rio), Rua Marquês de São Vicente 225, Rio de Janeiro, RJ – zip code 22453-900 – Brazil

[2] Associate Professor – Dept. of Structural and Foundations Engineering, Polytechnic School at the University of São Paulo (EPUSP), PO Box 61548 – São Paulo, SP – zip code 05424-970 – Brazil

Introduction

The fatigue crack propagation life prediction under complex loading in intricate two-dimensional (2D) structural components is a quite interesting modeling problem, requiring a mixed approach to achieve its optimum solution.

To predict the crack path and to calculate its associated stress intensity factors K_I and K_{II}, a finite element (FE) global discretization of the component, using appropriate crack tip elements, mesh generation schemes and crack increment criteria, has become a common engineering design practice. However, such brute force numerical calculation is not efficient when the load is complex, causing in the general case different crack increments at each load cycle, requiring remeshing and time-consuming recalculations in FE. Moreover, crack retardation effects compromise even more the computational efficiency of this approach.

On the other hand, the local approach, based on the direct integration of the crack propagation rule, can be efficiently used to calculate the crack increment at each load cycle, considering crack retardation effects if necessary. However, it requires as input the stress intensity expression for the crack, which is a major drawback because it is simply not available for most real components. Therefore, designers must use engineering common sense to choose approximate K_I handbook expressions to solve real problems. The error involved in such approximations obviously increases as the real crack deviates from the modeled crack, and in such cases the accuracy of the local approach is questionable and its predictions unreliable.

Since the advantages of the two approaches are complementary, the problem can be successfully divided into two steps. First, an appropriate FE software can be used to calculate the (generally curved) crack path and its associated Mode I stress intensity factor $K_I(a)$ along the crack length a, under **simple** loading. Then, an analytical expression should be adjusted to the discrete $K_I(a)$ calculated values, to be used as input to a local approach software. Finally, the actual **complex** loading can be efficiently treated by the direct integration of the crack propagation rule, considering retardation effects if required.

This is a simple and evident idea, but it is easier said than implemented, since FE experts are normally not fatigue design engineers, and vice-versa. Moreover, to achieve an engineering tool status, a two-step system must be numerically reliable, simple-to-use, versatile (since there is no universally accepted design method), and, of course, economic. All these constraints put a significant burden on the programming effort.

The purpose of this paper is to describe the fundamentals of an integrated system composed of two complementary programs, designed to implement this two-step method. This system demonstrates that satisfactory fatigue life predictions under complex load for 2D structural components can now be obtained on a PC platform.

The paper is organized in several sections describing (i) the numerical procedures to compute stress intensity factors in arbitrary 2D geometries; (ii) the principles of the crack increment direction numerical computation problem; (iii) a summary of the main features of the FE software developed to handle these tasks, named **QUEBRA2D**; (iv) an experimental verification of the predictions made by this software; (v) the fundamentals of the local approach to predict fatigue life under complex load; (vi) the main features of the **ViDa** software developed to perform such predictions; (vii) the fundamentals of the ΔK_{rms} method numerical implementation for 1D and 2D crack growth; (viii) the principles of the cycle-by-cycle method numerical implementation for 1D and 2D crack propagation; (ix) the minimum number of features required for modeling the load cycle interaction problems; and finally (x) a conclusion.

Numerical Computation of Stress-Intensity Factors

In 2D finite element models, three methods can be chosen to compute the stress-intensity factors along the (generally curved) crack path: the displacement correlation technique (*DCT*) [*1*], the potential energy release rate computed by means of a modified crack-closure integral technique (*MCC*) [*2, 3*], and the *J*-integral computed by means of the equivalent domain integral (*EDI*) together with a mode decomposition scheme [*4-8*].

Displacement Correlation Technique (DCT)

In *DCT*, the displacements obtained from the finite element analysis at specific locations are correlated with the analytic solutions expressed in terms of the stress-intensity factors. For quarter-point singular elements [*1*], the crack opening displacement δ is given by

$$\delta(r) = \left(4v_{j-1} - v_{j-2}\right)\sqrt{\frac{r}{L}} \tag{1}$$

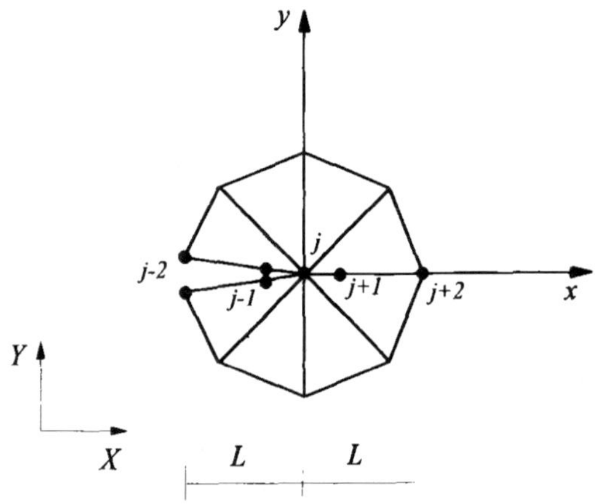

Figure 1 – *Quarter-point elements at the crack tip.*

where v_{j-1} and v_{j-2} are the relative displacements in the y direction, at the j–1 and j–2 nodes, and L is the element size (Figure 1). The analytical expression for δ is

$$\delta(r) = K_I\left(\frac{\kappa+1}{\mu}\right)\sqrt{\frac{r}{2\pi}} \tag{2}$$

where $\kappa = 3 - 4\nu$ for plane strain, $\kappa = (3 - \nu)/(1 + \nu)$ for plane stress, ν is the Poisson ratio, and μ is the shear modulus. By equating the numerical and the analytical expressions for δ, (1) and (2), the Mode I stress-intensity factor can be evaluated by

$$K_I = \left(\frac{\mu}{\kappa + 1}\right)\sqrt{\frac{2\pi}{L}}\left(4v_{j-1} - v_{j-2}\right) \tag{3}$$

For Mode II, the crack opening displacement is replaced by the crack sliding displacement and, following the same steps, K_{II} is calculated by

$$K_{II} = \left(\frac{\mu}{\kappa + 1}\right)\sqrt{\frac{2\pi}{L}}\left(4u_{j-1} - u_{j-2}\right) \tag{4}$$

where u_{j-1} and u_{j-2} are the relative displacements in the x direction, at the j–1 and j–2 nodes (Figure 1).

Modified Crack-Closure Integral (MCC)

The modified crack-closure method is based on the Irwin's crack-closure integral concept, which assumes that the required work to close a crack from $a + \delta a$ to a is the same as that required to extend it from a to $a + \delta a$ (Figure 2). Based on this assumption, the strain-energy release rates G_I and G_{II} of a mixed-mode condition are obtained by

$$G_I = \lim_{\delta a \to 0} \frac{1}{2\delta a}\int_0^{\delta a} v(r)\sigma_y(r)\,dr \quad \text{and} \quad G_{II} = \lim_{\delta a \to 0} \frac{1}{2\delta a}\int_0^{\delta a} u(r)\sigma_{xy}(r)\,dr \tag{5}$$

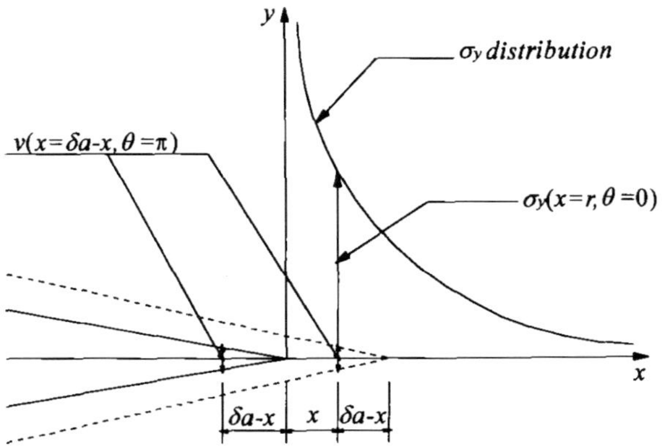

Figure 2 – *Analytical crack-closure integral method.*

In these equations, δa is the virtual crack extension; σ_y and σ_{xy} are the normal and shear stress distributions ahead of the crack tip; and $v(r)$ and $u(r)$ are the crack opening and sliding displacements at a distance r behind the new crack tip. In the original form, the results are obtained from two analyses: one with a crack length a and the other with a crack length $a + \delta a$.

Rybicki and Kanninen [2] were the first to use this approach with a single finite element analysis, using models with four-noded quadrilateral elements. Raju [3] extended this method for non-singular and singular elements of any order. This procedure is based on the symmetry of the elements around the crack tip. In the numerical computation of G_I and G_{II}, the strain-energy release rates given by Equation 5, the stress field is assumed to have the classical $1/\sqrt{r}$ distribution and the displacements $u(r)$ and $v(r)$ are determined by interpolation of nodal displacements using the element shape functions. The normal and shear stresses are obtained from the nodal forces at and ahead of the crack tip.

As shown by Raju [3], simplified expressions for singularity elements may be applied, which are easier to use than the consistent expressions. The components G_I and G_{II} for pure Mode I and Mode II, and for mixed mode conditions are given as

$$G_I = -\frac{1}{2\delta a}\left[F_{y_i}\{t_{11}(v_m - v_{m'}) + t_{12}(v_l - v_{l'})\} + F_{y_j}\{t_{21}(v_m - v_{m'}) + t_{22}(v_l - v_{l'})\}\right] \quad (6)$$

$$G_{II} = -\frac{1}{2\delta a}\left[F_{x_i}\{t_{11}(u_m - u_{m'}) + t_{12}(u_l - u_{l'})\} + F_{x_j}\{t_{21}(u_m - u_{m'}) + t_{22}(u_l - u_{l'})\}\right] \quad (7)$$

where F_{x_i}, F_{x_j}, F_{y_i}, and F_{y_j} are the consistent nodal forces acting on nodes i and j in the x and y directions (Figure 3); u and v are the nodal displacements at m, m', l and l' nodes in the x and y directions, respectively; and $t_{11} = 6 - 3\pi/2$, $t_{12} = 6\pi - 20$, $t_{21} = 1/2$, and $t_{22} = 1$. The nodes and nodal forces F_{y_i} and F_{y_j} are shown in Figure 3.

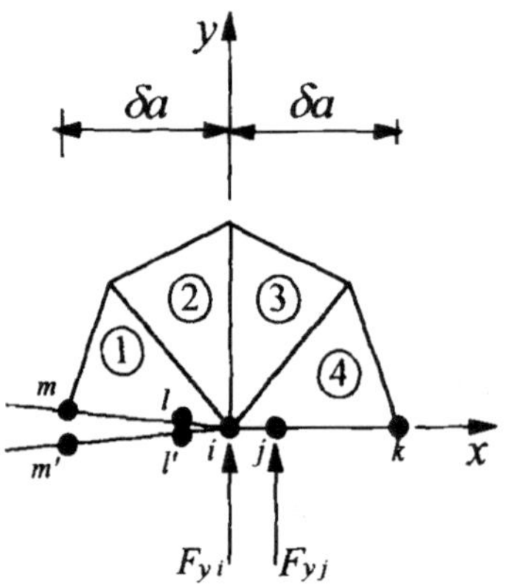

Figure 3 – *Node at crack tip elements and consistent nodal forces ahead of crack tip.*

The nodal forces F_{x_i} and F_{y_i} are computed from elements 1, 2, 3 and 4, but the forces F_{x_j} and F_{y_j} are computed from element 4 only. Under linear elastic conditions (LEFM), the stress-intensity factors are related to the energy release rates by

$$G_I = \frac{\kappa + 1}{8\mu} K_I^2 \quad \text{and} \quad G_{II} = \frac{\kappa + 1}{8\mu} K_{II}^2 \tag{8}$$

where κ has been previously defined for plane stress and plane strain conditions. It is assumed that the classical ASTM E399 requirements for validating a $\mathbf{K_{IC}}$ toughness test can also be used in fatigue crack growth to characterize a plane strain condition.

J-Integral Formulation with Equivalent Domain Integral (EDI)

The *J*-integral is a path independent contour integral introduced by Rice [9] to study non-linear elastic materials under small scale yielding. The equivalent domain integral method replaces the integration along the contour by another one over a finite size domain, using the divergence theorem. This definition is more convenient for finite element analysis. For two-dimensional problems, the contour integral is replaced by an area integral

$$J = -\int_A \left[W \frac{\partial q}{\partial x} - \sigma_{ij} \frac{\partial u_i}{\partial x} \frac{\partial q}{\partial x} \right] dA - \int_A \left\{ \frac{\partial W}{\partial x} - \frac{\partial}{\partial x}\left[\sigma_{ij} \frac{\partial u_i}{\partial x} \right] \right\} q dA - \int_S t_i \frac{\partial u_i}{\partial x} q ds \tag{9}$$

where W is the strain energy density; q is a continuous function that allows for the equivalent domain integral to be treated in the finite element formulation; σ_{ij} are the stresses at the contour C, which is any path surrounding the crack tip; u_i are the displacements correspondent to local i-axes; t_i is the crack face pressure load, and s is the arc length of the contour. Usually, a linear function is chosen for q, which assumes a unit value at the crack tip and a null value along the contour. For the linear-elastic materials special case, the second term in Equation 9 vanishes. The third term will vanish if the crack faces are not loaded, or if $q = 0$ at the loaded portions of the crack faces.

The *J*-integral definition considers a balance of mechanical energy for a virtual translation field along the x-axis. In the case of either pure Mode I or pure Mode II, Equation 9 allows for the calculation of stress-intensity factors K_I or K_{II}. However, in the mixed mode case, K_I and K_{II} can not be calculated separately from this equation alone. In this case, other invariant integrals are used. Usually, the expression defined by Knowles and Sternberg [10] is adopted

$$J_k = -\int_A \left[W \frac{\partial q}{\partial x_k} - \sigma_{ij} \frac{\partial u_i}{\partial x_k} \frac{\partial q}{\partial x_j} \right] dA - \int_A \left[\frac{\partial W}{\partial x_k} - \sigma_{ij} \frac{\partial}{\partial x_j}\left(\frac{\partial u_i}{\partial x_k} \right) \right] q dA - \int_S t_i \frac{\partial u_i}{\partial x_k} q ds \tag{10}$$

where k is an index for local crack tip axes (x, y). These integrals were introduced initially for small deformation [9] and were extended by Atluri [11] for finite deformations.

The integration is performed in the elements chosen to represent the domain. In this work, the chosen domain is the rosette of quarter-point elements at the crack tip (Figure 1), and the standard Gaussian quadrature is used over each element.

For linear elastic problems, Bui [4] proposed associated fields to decompose the loading modes. In this case, the first component in Equation 10 is path independent, but the second one is not. However, the path dependency may be eliminated if the displacements and the stress fields are decomposed into symmetric and anti-symmetric portions. Therefore, the displacement field is rewritten as

$$\begin{aligned} u &= u^{I} + u^{II} = \frac{1}{2}(u + u') + \frac{1}{2}(u - u') \\ v &= v^{I} + v^{II} = \frac{1}{2}(v - v') + \frac{1}{2}(v + v') \end{aligned} \tag{11}$$

where u and v are displacements in x and y directions, respectively, $u'(x,y) = u'(x,-y)$, and $v'(x,y) = v'(x,-y)$; and the superscripts I and II correspond to the symmetric and anti-symmetric components of the displacements field, respectively. The stress field is then decomposed as

$$\begin{aligned} \sigma_{xx} &= \sigma_{xx}^{I} + \sigma_{xx}^{II} = \frac{1}{2}(\sigma_{xx} + \sigma'_{xx}) + \frac{1}{2}(\sigma_{xx} - \sigma'_{xx}) \\ \sigma_{yy} &= \sigma_{yy}^{I} + \sigma_{yy}^{II} = \frac{1}{2}(\sigma_{yy} + \sigma'_{yy}) + \frac{1}{2}(\sigma_{yy} - \sigma'_{yy}) \\ \sigma_{zz} &= \sigma_{zz}^{I} + \sigma_{zz}^{II} = \frac{1}{2}(\sigma_{zz} + \sigma'_{zz}) \\ \sigma_{xy} &= \sigma_{xy}^{I} + \sigma_{xy}^{II} = \frac{1}{2}(\sigma_{xy} - \sigma'_{xy}) + \frac{1}{2}(\sigma_{xy} + \sigma'_{xy}) \end{aligned} \tag{12}$$

where $\sigma'_{ij}(x,y) = \sigma'_{ij}(x,-y)$ and $\sigma_{zz}^{II} = 0$.

New integrals J_I and J_{II} are obtained, which satisfy the condition $J = J_I + J_{II}$, where J_I is associated to the symmetric field (Mode I) and J_{II} is associated to anti-symmetric fields (Mode II)

$$J_I = -\int_A \left[W(u_i^I)\frac{\partial q}{\partial x_k} - \sigma_{ij}(u_i^I)\frac{\partial u_i^I}{\partial x_k}\frac{\partial q}{\partial x_j} \right] dA - \int_S t_i \frac{\partial u_i^I}{\partial x_k} q\, ds \tag{13}$$

$$J_{II} = -\int_A \left[W(u_i^{II})\frac{\partial q}{\partial x_k} - \sigma_{ij}(u_i^{II})\frac{\partial u_i^{II}}{\partial x_k}\frac{\partial q}{\partial x_j} \right] dA - \int_S t_i \frac{\partial u_i^{II}}{\partial x_k} q\, ds \tag{14}$$

This approach has also been applied by Atluri et al. [7, 8] with highly accurate results for mixed-mode problems. In addition, Eischen [12], and Kienzler and Kordisch [13] suggested improved methods for obtaining J-integrals for mixed-mode problems. These modifications and decomposition techniques permit the use of the J-integral and EDI approaches for a wide range of linear and non-linear deformation crack problems.

In LEFM, J is equal to the energy release rate G, and its components J_I and J_{II} may be used to compute stress-intensity factors by means of Equation 8.

Numerical Computation of the Crack Increment Direction

In 2D finite element analysis, the three most used criteria for numeric computation of crack (incremental) growth in the linear-elastic regime are: (a) the maximum circumferential stress ($\sigma_{\theta max}$) [*14*], (b) the maximum potential energy release rate ($G_{\theta max}$) [*15*], and (c) the minimum strain energy density ($S_{\theta min}$) [*16*].

In the first criterion, Erdogan and Sih considered that the crack extension should occur in the direction that maximizes the circumferential stress in the region close to the crack tip [*14*]. In the second, Hussain et al. [*15*] have suggested that the crack extension occurs in the direction that causes the maximum fracturing energy release rate. And in the last, Sih [*16*] assumed that the crack growth direction is determined by the minimum strain energy density value near the crack tip. Bittencourt et al. [*17*] have shown that, if the crack orientation is allowed to change in automatic fracture simulation, the three criteria furnish basically the same results. Since the maximum circumferential stress criterion is the simplest, it is the criterion described below.

The stresses on the crack tip for Modes I and II are given by summing up the stresses obtained for each mode separately [*18*]. As a result, the following equations are obtained in polar coordinates

$$\sigma_r = \frac{1}{\sqrt{2\pi r}} cos(\theta/2) \left\{ K_I \left[1 + sin^2(\theta/2)\right] + \tfrac{3}{2} K_{II}\, sin\theta - 2K_{II} tan(\theta/2) \right\} \quad (15)$$

$$\sigma_\theta = \frac{1}{\sqrt{2\pi r}} cos(\theta/2) \left[K_I cos^2(\theta/2) - \tfrac{3}{2} K_{II}\, sin\theta \right] \quad (16)$$

$$\tau_{r\theta} = \frac{1}{\sqrt{2\pi r}} cos(\theta/2) \left[K_I\, sin\theta + K_{II}(3cos\theta - 1)\right] \quad (17)$$

These expressions are valid both for plane stress and plane strain. The maximum circumferential stress criterion determines that the crack extension begins on a plane perpendicular to the direction in which σ_θ is maximum, thus $\tau_{r\theta} = 0$, and that the monotonic (non-fatigued) extension shall occur when $\sigma_{\theta max}$ reaches a critical value corresponding to a property of the material (K_{IC} for Mode I). From Equations 15-17 and $\tau_{r\theta} = 0$, it is found a trivial solution $\theta = \pm\pi$ for $cos(\theta/2) = 0$, and a non-trivial solution

$$K_I sin\theta + K_{II}(3cos\theta - 1) = 0 \quad (18)$$

Analyzing Equation 18 for the two pure modes, it is found for pure Mode I that $K_{II} = 0$, $K_I\, sin\theta = 0$, and $\theta = 0°$, and for pure Mode II that $K_I = 0$, $K_{II}\,(3cos\theta - 1) = 0$, and $\theta = \pm 75°$. These θ values are the extreme values of the crack propagation angle. The intermediate values are found solving Equation 18 for θ considering the mixed mode

$$\theta = 2\,arctan\left(\frac{1}{4}\frac{K_I}{K_{II}} \pm \frac{1}{4}\sqrt{\left(\frac{K_I}{K_{II}}\right)^2 + 8} \right) \quad (19)$$

where the sign of θ is the opposite of the sign of K_{II}.

Finite Element Crack Propagation Simulation

The computational models described above were implemented in a software called **QUEBRA2D** (meaning 2D fracture in Portuguese) [*19, 20*], which is an interactive graphical software for simulating two-dimensional fracture processes based on a finite element adaptive mesh generation strategy [*21*]. The adaptive process first requires the results from the analysis of an initial finite element mesh, usually rough, with the geometric descriptions, the boundary conditions, and their attributes. Then a discretization of the domain's region boundary is performed based on the geometric properties and on the characteristic sizes of the boundary elements, determined from the error estimate resulting from the previous step FE analysis.

It is important to point out that one advantage of this strategy is that the boundary curve is discretized independently from the model's domain, thus resulting in a more regular boundary discretization. From this discretization, the new mesh is generated [*22*], based on quadtree [*23*] and Delaunay [*24*] triangulation techniques. The quadtree generates the mesh in the interior of the model, leaving a band near the boundary to be discretized by the Delaunay triangulation. This process is repeated until the estimate discretization error reaches a predefined value [*19, 21*].

Some other **QUEBRA2D** highlights are: (i) visualization of iso-strips and iso-lines from scalar results at the nodes and at the Gauss points; (ii) stress-intensity factor and crack propagation direction computation by means of all methods described above; (iii) vectorial or scalar plotting for visualizing the principal stress results; (iv) visualization of the model's deformed configuration, with zoom, distortion, and translation specification; (v) visualization of the model's animation along the several steps; and (vi) option of the interface language.

The software has been implemented in C language, using the IUP/LUA interface system (http://www.tecgraf.puc-rio.br/manual/iup) and the CD graphic system (http://www.tecgraf.puc-rio.br/manual/cd). This environment allows, without any code modification, automatic portability to several platforms, including workstations based on the Unix operating system and PCs running under Windows 98/2000 or NT.

Experimental Verification of the Crack Path Prediction

A simple test was performed to verify the crack path predicted by the **QUEBRA2D** software and to demonstrate its capabilities. A crack was fatigue propagated in a SEN specimen with a hole slightly to the left of the starting notch (created using a 0.3mm jeweler's saw), loaded in four-point bending (Figure 4). Due to the hole, the crack does not follow a straight line path, but curves toward the hole.

The material used in the experiment was a 1020 steel (analyzed composition: C 0.19; Mn 0.46; Si 0.14; Ni 0.052; Cr 0.045; Mo 0.007; Cu 0.11; Nb 0.002; Ti 0.002; Fe balance) with Young modulus $E = 205GPa$, yielding strength $S_Y = 285MPa$, ultimate strength $S_U = 491MPa$, and area reduction $RA = 54\%$. Two crack growth equations were fitted to experimental data (obtained testing CTS in a servohydraulic test machine under sinusoidal load at 20Hz): the Paris equation, yielding $da/dN = 8.5 \cdot 10^{-14} \cdot \Delta K^{4.2}$, and a modified Elber equation, $da/dN = 4.5 \cdot 10^{-10} \cdot (\Delta K - \Delta K_{th})^{2.1}$, where the threshold stress intensity range $\Delta K_{th} = 11.6MPa\sqrt{m}$.

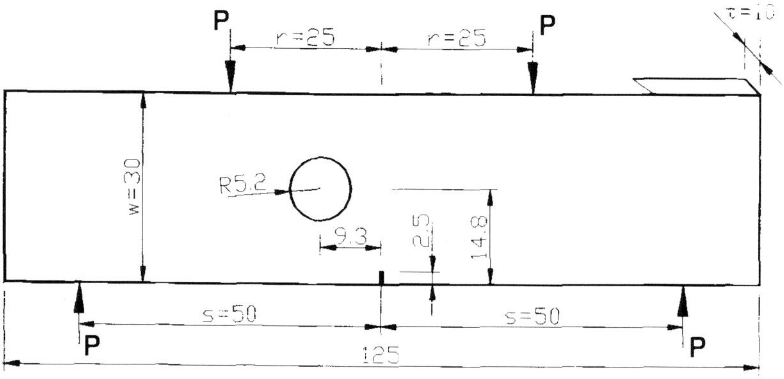

Figure 4 – *SEN specimen with a hole to the left of the starting notch (dimensions in mm).*

The experiment was performed under constant stress intensity range $\Delta K=18.0MPa\sqrt{m}$ and $R = 0.1$. The maximum loads varied from $11.0kN$ to $2.8kN$ to keep ΔK constant as the crack propagated from $2.5mm$ to $16.5mm$ (measured along the crack path).

Figure 5 shows the FE mesh automatically generated for the final crack configuration. Figure 6 compares the predicted crack path with the actual one. The crack path was predicted by the $\sigma_{\theta max}$ method and K_I was computed by the MCC technique. For refined meshes such as the one shown in Figure 5, all methods predict essentially the same results, as discussed in [*17*]. Finally Figure 7 presents the calculated Mode I stress intensity factor along the crack path, $K_I(a)$. This is the information that the local approach uses for calculating the fatigue life under complex loading.

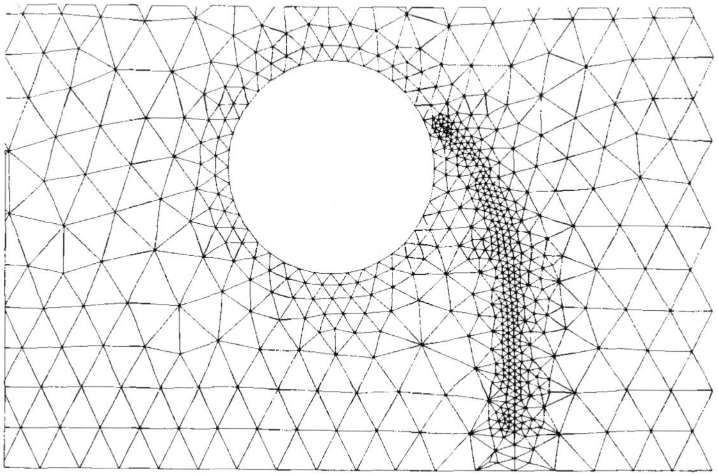

Figure 5 – *FE mesh automatically generated for the final crack configuration.*

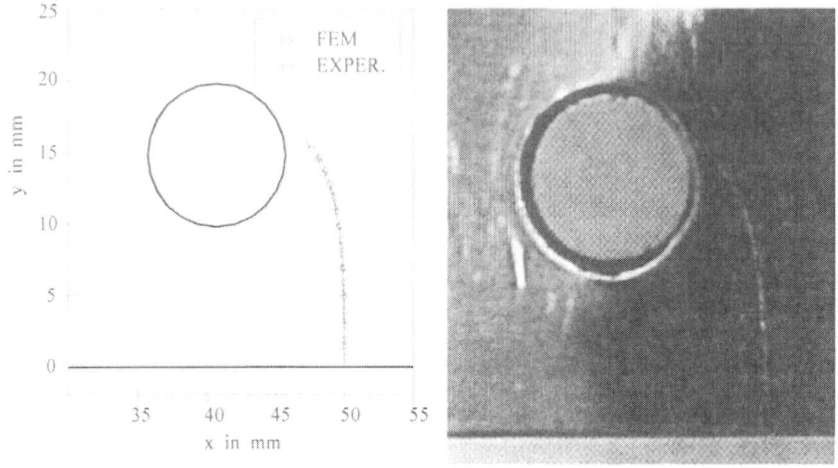

Figure 6 – *Predicted and measured crack path.*

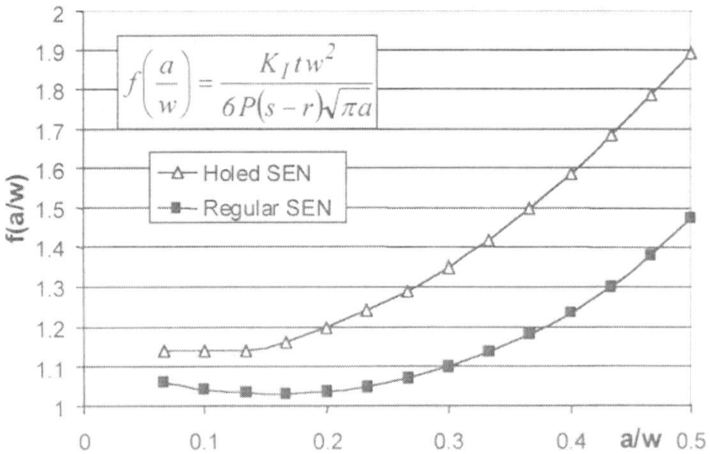

Figure 7 – *Calculated $K_I(a)$ along the crack path for the holed and regular SEN.*

Automation of the Fatigue Crack Propagation Calculation under Complex Loading

The modeling and the calculation automation of the LEFM Mode I fatigue crack propagation under complex loading by the **local** approach are discussed below. Both 1D and 2D cracks are studied, even though these last are not dealt with in the FE modeling discussed above. The loading complexity, whose amplitude can randomly vary in time, is not limited. Sequence effects, such as overload-induced crack retardation or arrest are also considered. Only Mode I is discussed, since fatigue cracks almost always propagate perpendicular to the maximum tensile stress.

The local approach is so-called because it does not require the global solution of the structure's stress field, because it is based on the direct integration of the fatigue crack propagation rule of the material, $da/dN = F(\Delta K, R, \Delta K_{th}, K_C, ...)$, where ΔK is the stress intensity range, $R = K_{min}/K_{max}$ is a measure of the mean load, ΔK_{th} is the fatigue crack propagation threshold, and K_C is the toughness of the material-structure. Appropriate stress intensity factor expression for ΔK and da/dN rule must be used to obtain satisfactory predictions. Therefore, neither the ΔK expression nor the type of crack propagation rule should have their accuracy compromised when using this approach.

The interaction with the environment and out-of-phase loading at multiple origins, which induce stresses whose principal directions vary significantly in time, are considered out of the scope of this discussion. In the same way, it does not consider the small crack problem, whose size is of the order of (i) the size parameter that characterizes the intrinsic anisotropy of the material (e.g., grain size), or (ii) the plastic zone associated to the crack tip or to the notch in which the (short) crack is built-in.

In the sequence of this text, first the main features of the software **ViDa** (which means life in Portuguese) are concisely described. This software has been developed to automate all the traditional local approach methods used in fatigue design [*25, 26*], including the *S-N*, the *IIW* (for welded structures) and the ε-*N* for crack initiation, and the *da/dN* for crack propagation. Then the following topics are discussed: (i) the ΔK_{rms} method, including the differences between 1D and 2D crack propagation modeling; (ii) the cycle-by-cycle method, also emphasizing the differences between the 1D and the 2D problems; (iii) some proposals for increasing the computational efficiency of the models; and (iv) the modeling of load sequence effects. Finally, the advantages and limitations of the several studied models are evaluated.

The ViDa Software

The objective of this software is to automate all the calculations required to predict fatigue life under complex loading using the local approach. It runs on PCs under Windows 95 or better operating system, and it includes all the necessary tools to perform the predictions, such as intuitive and friendly graphic interfaces in multiple idioms; intelligent databases for stress concentration and intensity factors, crack propagation rules, material properties, and the like; traditional and sequential rain-flow counters, graphic generators of elastoplastic hysteresis loops and of 2D cracks fronts; automatic adjustment of crack initiation and propagation experimental data; an equation interpreter, etc. The software calculates crack growth considering any propagation rule and any ΔK expression that can be written in a BASIC syntax (making it an ideal companion to **QUEBRA2D** software, which can be used to generate the $\Delta K(a)$ expression if it is not available in its database).

The loading can be given by a sequential list of peaks (σ_{max}) and valleys (σ_{min}), or else by the equivalent sequence of the number of reversions ($n/2$) of alternate (σ_a) and mean (σ_m) stresses. The loading can also be specified in strain instead of stress. The data can be typed or imported from any text file, including those experimentally generated (e.g. by strain-gages).

The propagation is calculated at each load event. An event is defined by a block of simple load, in which σ_a and σ_m remain constant during *n* cycles, or at each variation of the load amplitude in the complex case. In any case, the software automatically stops the calculations, and indicates the value of the parameters that caused the stop, if during the loading it detects that: (i) $K_{max} = K_C$; (ii) the crack reaches its maximum specified size; (iii) the stress in the residual ligament reaches the rupture strength of the material S_U; (iv) the crack propagation rate *da/dN* reaches 0.1mm/cycle (above this rate fracturing occurs, not fatigue cracking); or (v) one of the borders of the piece is reached by the front of the crack, in the 2D crack propagation case. However, for some geometries, the software is able to calculate 2D crack propagation even after the borders of the piece are reached, by modeling the stress intensity factors of the transition from part-through to through cracks.

Moreover, the software informs when there is yielding in the residual ligament before the maximum specified crack size or number of load cycles is reached. In this way, the calculated values can be used with the guarantee that the limit of validity of the mathematical models is never exceeded.

The ΔK_{rms} Method

The stress intensity factor range is expressed as $\Delta K = \Delta\sigma\cdot[\sqrt{(\pi a)}\cdot f(a/W)]$, where $\Delta\sigma$ is the nominal stress range (in relation to which the ΔK expression is defined), *a* is the crack length, $f(a/W)$ is a non-dimensional function of a/W, and *W* is a characteristic size of the structure. Therefore, $\Delta\sigma$ quantifies the influence of the loading and $\sqrt{(\pi a)}\cdot f(a/W)$ quantifies the effect of the geometry of the piece and of the crack shape and size in ΔK.

The simplest way to treat the fatigue life prediction under a complex loading problem is to substitute a simple equivalent loading, causing the same growth of the crack. It has been experimentally discovered that ΔK_{rms}, the root mean square value of the stress intensity range, can in some cases be used for this purpose [*18*].

According to Hudson [*27*], ΔK_{rms} can be calculated from the *rms* values of the positive peaks and valleys of the loading (since the crack does not grow while closed, the compressive part of the loading should be discarded). Therefore

$$\sigma_{max_{rms}} = \sqrt{\frac{1}{p}\sum_{i=1}^{p}(\sigma_{max_i})^2} \quad \text{and} \quad \sigma_{min_{rms}} = \sqrt{\frac{1}{q}\sum_{i=1}^{q}(\sigma_{min_i})^2}\ , \quad (\sigma_{max_i}, \sigma_{min_i} \geq 0) \qquad (20)$$

$$\Delta\sigma_{rms} = \sigma_{max_{rms}} - \sigma_{min_{rms}} \quad \text{and} \quad R_{rms} = \frac{\sigma_{min_{rms}}}{\sigma_{max_{rms}}} \qquad (21)$$

As $\Delta K_{rms} = \Delta\sigma_{rms}\cdot[\sqrt{(\pi a)}\cdot f(a/W)]$, the number of cycles the crack takes to grow from the initial length a_0 to the final one a_f is given by

$$N = \int_{a_0}^{a_f} \frac{da}{F(\Delta K_{rms}, R_{rms}, \Delta K_{th}, K_C, \ldots)} \qquad (22)$$

In the **ViDa** software, a variation of the Simpson's algorithm can be used for the numerical integration of the simple loading case and, consequently, also for the ΔK_{rms} method. Crack increments $\delta a > 0.1\mu m$ for the discretization of the integral can be specified by the user, who can also choose an integration method based on adjustable steps depending on the variation of the crack length, as will be discussed later on the study of the cycle-by-cycle method.

It should be mentioned that the ΔK_{rms} value of a complex loading is similar but not identical to the ΔK of a simple loading. As with any statistics, ΔK_{rms} does not recognize temporal order, and cannot detect some important problems such as: sudden fracture caused by a single large peak during the complex loading (in order to start the fracture process, it is enough that in just **one** event $K_{max} = K_C$); or any interaction among the loading cycles (e.g. the crack retardation or arrest phenomena after an overload). Also, it is **not** possible to guarantee the inactivity of the crack if $\Delta K_{rms}(a_0) < \Delta K_{th}(R_{rms})$.

In complex loading, this latter problem can be caused by all the ($\Delta\sigma_i$, R_i) events that induce $\Delta K_i > \Delta K_{th}(R_i)$, which can make the crack grow even if $\Delta K_{rms} < \Delta K_{th}(R_{rms})$. Therefore, as ΔK_i depends both on the stress range $\Delta\sigma_i$ and on the crack size a_i in that event, even if the value of $\Delta\sigma_{rms}$ stays constant, the same cannot be guaranteed for ΔK_{rms}.

The ΔK_{rms} Method for 2D Cracks

Equation 22 can only be applied to 1D cracks, but in practice many times it is necessary to study surface, corner, or internal cracks that propagate in 2D. The principal characteristic of these cracks is a non-homologous fatigue propagation: in general, the crack front tends to change form from cycle to cycle, because ΔK varies from point to point along the crack front.

There are analytical expressions for the stress intensity factor of some 2D cracks. If the cracks have ellipsoidal fronts built in a plate of width W or $2W$ and thickness t, ΔK is function of $\Delta\sigma$, a, a/c, a/t, c/W and θ *[28-30]*, where a and c are the ellipsis semi-axes, and θ is defined in Figure 8.

The 2D ellipsoidal crack propagation problem is a reasonable approximation for many actual surface, corner, or internal cracks. Fractographic observations indicate that the successive fronts of those cracks tend to achieve an elliptical form, see Figure 9, and to stay approximately elliptic during their fatigue propagation, even when the initial crack shape is far from an ellipsis *[31, 32]*. Therefore, it can be quite reasonable to assume in the modeling that the fatigue propagation just changes the shape of the 2D cracks (given by the ratio a/c between the ellipsis semi-axes, which quantifies how elongated the cracks are), but preserves their basic ellipsoidal geometry.

As an ellipsis is completely defined by its two semi-axes, to predict the growth of 2D (elliptical) cracks, including their shape changes, it is enough to calculate at each load cycle the lengths of the ellipsis axes a and c, **jointly** solving the da/dN and the dc/dN propagation problems.

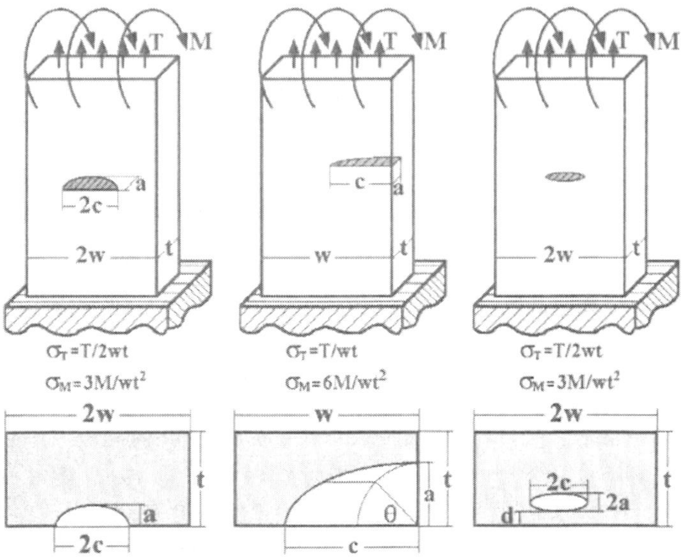

Figure 8 – *Surface semi-elliptical, corner quart-elliptical, and internal elliptical cracks.*

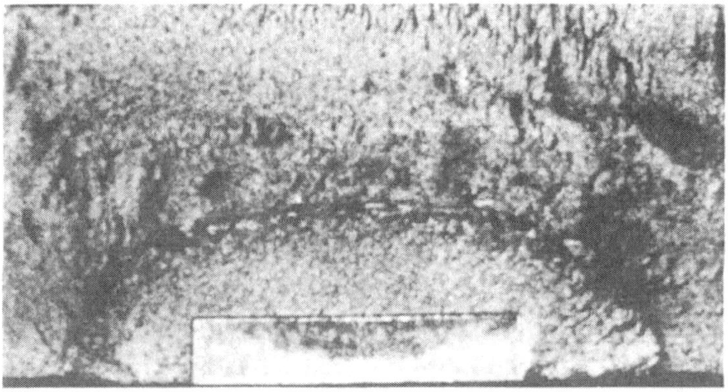

Figure 9 – *A surface fatigue crack that started from a sharp rectangular notch and grew with an approximately semi-elliptical front.*

To exemplify the simplest case of this type of problem, the semi-elliptical surface crack with $a < c$ under pure normal loading $\Delta\sigma$ is briefly analyzed. The expression for $\Delta K(a) = \Delta\sigma \cdot [\sqrt{(\pi a)} \cdot f_a(a/c,\ a/t,\ c/W)]$ from the Newman and Raju solution [*29*] is quite complex, although the ratio $\Delta K(a)/\Delta K(c)$ is relatively simple, allowing for the visualization of the basic ideas of the calculation methodology

$$\Delta K(c) = \Delta K(a) \cdot \sqrt{\frac{a}{c}} \cdot \left[1.1 + 0.35(a/t)^2\right] \tag{23}$$

The calculation model requires the crack type and initial size a_0 and c_0, the bar geometry, the loading type, the mechanical properties, and the crack propagation rule $da/dN = F(\Delta K, R, \Delta K_{th}, K_c, ...)$ of the material. A small crack increment δa must also be specified (50μm, a number of the order of the resolution threshold of the crack measurement methods in fatigue tests, can be a good choice both from the physical and from the numeric points of view). From the complex loading, $\Delta\sigma_{rms}$ and R_{rms} are calculated to obtain

$$\Delta K_{rms}(a_0) = \Delta\sigma_{rms} \cdot [\sqrt{(\pi a_0)} \cdot f_a(a_0/c_0, a_0/t, c_0/W)] \tag{24}$$

The number of cycles N_0 the crack takes to grow from a_0 to $a_0 + \delta a$ is given by

$$N_0 = \frac{\delta a}{F(\Delta K_{rms}(a_0), R_{rms}, \Delta K_{th}, K_c, ...)} \tag{25}$$

$\Delta K_{rms}(c_0)$ can be calculated by Equation 23, or by an expression similar to Equation 24, to get the correspondent growth in the direction of the semi-axis c, δc_0, which is given by

$$\delta c_0 = N_0 \cdot F(\Delta K_{rms}(c_0), R_{rms}, \Delta K_{th}, K_c, ...) \tag{26}$$

The numerical calculation process can now start the coupled interactions making $\Delta K_{rms}(a_1) = \Delta K_{rms}(a_0 + \delta a)$ and calculating $N_1 = \delta a/F(\Delta K_{rms}(a_1), R_{rms}, \Delta K_{th}, K_c, ...)$, in order to get $\delta c_1 = N_1 \cdot F(\Delta K_{rms}(c_1), R_{rms}, \Delta K_{th}, K_c, ...)$, where $c_1 = c_0 + \delta c_0$, etc. The precision of the methodology can be adjusted by the value of δa.

When compared to the 1D growth, the application of the ΔK_{rms} method to the 2D problem is more laborious and uses a less efficient integration method but does not present supplementary conceptual difficulties. However, it should be noticed that the 2D propagation presents some particularities that differentiate it from the 1D case. As these cracks have different values for $\Delta K(a)$ and $\Delta K(c)$, there are three distinct 2D propagation cases under simple loading (assuming $\Delta K(a) > \Delta K(c)$ to start with)

1) $\Delta K(a_0)$ and $\Delta K(c_0) > \Delta K_{th}$: the crack spreads in both directions, changing shape at each i-th load cycle depending on the ratio $\Delta K(a_i)/\Delta K(c_i)$.
2) $\Delta K(a_0) > \Delta K_{th}$ and $\Delta K(c_0) < \Delta K_{th}$: the crack grows only in the a direction, until its size is big enough to make $\Delta K(c_0) > \Delta K_{th}$, when the problem reverts to Case 1 (there are, however, pathological cases in which $\Delta K(a)$ **decreases** with a, and in these cases a crack can start spreading to later on stop if it reaches $\Delta K(a) < \Delta K_{th}$). Moreover, it is worth mentioning that this Case 2 is particularly deceiving in inspections, since the trace of a surface crack can remain constant during millions of cycles [*32*], apparently hinting that it is inactive, when in fact it is growing toward the inside of the piece, until reaching $\Delta K(c_0) > \Delta K_{th}$, when it starts to spread laterally in a relatively fast rate.
3) $\Delta K(a_0)$ and $\Delta K(c_0) < \Delta K_{th}$: the crack does not propagate.

In the complex loading case there are other details to consider. For instance, it is not possible to guarantee 2D crack inactivity if $\Delta K_{rms}(a_0)$ and $\Delta K_{rms}(c_0) < \Delta K_{th}$. But, due to space limitations, the study of these details is left for another work.

To conclude, it is worthwhile remembering that the ΔK_{rms} method is the simplest way to treat a complex loading problem, but it should only be used, as with any model, with the due appreciation of its limitations.

The Cycle-by-Cycle Method

In the cycle-by-cycle method, each load reversion is associated to the crack growth it would cause if it was the only one to load the piece (this implicates neglecting interaction effects among the several events of a complex loading, such as overload-induced retardation or arrest in the crack growth). Using this assumption, it is easy to write a general expression for the cycle-by-cycle crack growth, by any crack propagation rule: if $da/dN = F(\Delta K, R, \Delta K_{th}, K_C, \ldots)$, and if in the i-th 1/2 cycle of the loading the length of the crack is a_i, the stress range is $\Delta\sigma_i$ and the mean load causes R_i, then the crack grows by a δa_i given by

$$\delta a_i = \frac{1}{2} \cdot F(\Delta K(\Delta\sigma_i, a_i), R(\Delta\sigma_i, \sigma_{max_i}), \Delta K_{th}, K_C \ldots) \tag{27}$$

The total growth of the crack is quantified by $\Sigma(\delta a_i)$. Therefore, the cycle-by-cycle rule is similar in concept to the linear damage accumulation used in the *SN* and εN fatigue design methods. And, as in Miner's rule, it requests that **all** the events that cause fatigue damage be recognized before the calculation, by rain-flow counting the loading.

However, this counting algorithm alters the **order** of the loading, as shown in Figure 10. This can cause serious problems in the predictions, because the loading order effects in crack propagation are of two different natures.

- Delayed effects can retard or stop the subsequent growth of the crack due, e.g., to plasticity-induced Elber-type crack closure [*33*] or to crack tip bifurcation. These interaction effects among the loading cycles normally increase the crack life and, if neglected in the calculation, may induce excessively conservative predictions.
- Instantaneous fracture occurs when $K_{max} = K_C$ in **one** event, which must be precisely predicted.

As already mentioned above, the loading input in the **ViDa** software is sequential, and preserves the time order information that is lost when histograms or any other loading statistics are generated. To take advantage of this feature, a **sequential** rain-flow counting option was introduced in the software (Figure 10). The sequential rain-flow reorders the results from the traditional rain-flow based on the ending point location of each counted range pair [*25*]. With this technique, the effect of each large loading event is counted when it happens (and not before its occurrence, as in the traditional rain-flow method).

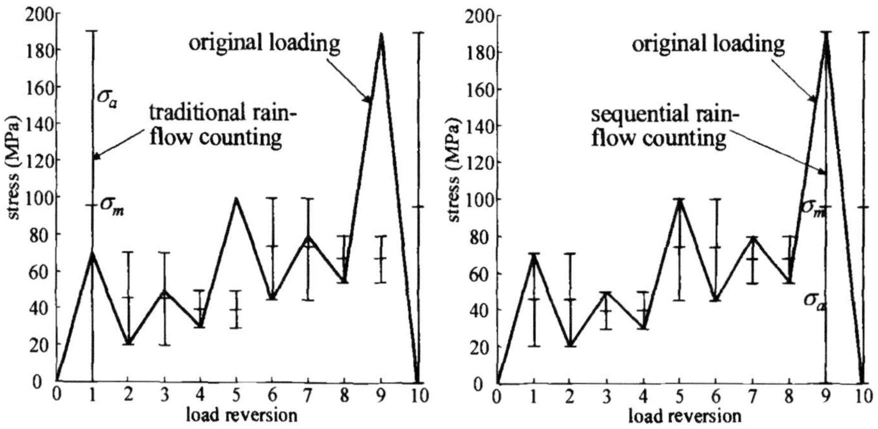

Figure 10 – *Traditional rain-flow counting, anticipating the large load events, and sequential rain-flow counting, which preserves most of the loading order.*

The main advantage of the sequential rain-flow counting algorithm is to avoid the premature calculation of the overload effects, which can cause **non**-conservative crack propagation life predictions (as $K(\sigma, a)$ in general grows with the crack, a given overload applied when the crack is large can be much more harmful than applied when the crack is small). The sequential rain-flow does not eliminate all the sequencing problems caused by the traditional method, but it is certainly an advisable option because it presents advantages over the original algorithm, without increasing its difficulty.

As discussed in the ΔK_{rms} method, the compressive part of the loading can be discarded in the calculations, that is, the negative peaks and valleys can be zeroed before the computations to decrease the numerical effort of the cycle-by-cycle method. And, in the same way, a range filtering option can be very useful to discard small loads that cause no damage inducing $\Delta K_i < \Delta K_{th}(R_i)$, following the ideas of the race-track method [*34*].

The range filtering can indeed significantly reduce the computational effort in fatigue damage calculations if the complex loading history is long. But this procedure is intrinsically **non**-conservative, because it can disregard damaging events as ΔK_i is not available before the crack growth calculations (ΔK depends not only on the loads, but also on the size of the crack). In consequence, the conservative rule is to limit the cut of the loading to the pairs $(\Delta\sigma_i, R_i)$ that cause $\Delta K(a_f) < \Delta K_{th}[R(a_f)]$, where a_f is the expected final length for the crack. But, in practice it can be easier numerically to try decreasing the ranges for the filtering, until there is no significant variation in the results.

The computational implementation of Equation 27, even with the pre-zeroing of the compressive peaks and valleys and with the range filtering of the loading, is still not numerically efficient. For this reason, an additional feature to reduce the computational time can be quite useful: the option of maintaining the geometrical part of ΔK constant during small variations in crack size.

As $\Delta K = \Delta\sigma \cdot [\sqrt{(\pi a)} \cdot f(a/W)]$, where $f(a/W)$ is a non-dimensional function (usually complicated) that depends only on the piece and crack geometry and not on the loading, it can be said that the range of the stress intensity factor ΔK_i at each load reversion depends on two variables of different nature:

1. on the stress range $\Delta\sigma_i$ in that event, and
2. on the length of the crack a_i in that instant.

$\Delta\sigma_i$, of course, can vary significantly at each load reversion when the loading is complex, but fatigue cracks always grow very slowly. In fact, at least in structural metals, the largest rates of stable crack growth observed in practice are of the order of μm/cycle, and during most of the life the crack growth rates are better measured in nm/cycle.

However, as in general the usually complicated $f(a/W)$ expressions do not present discontinuities, one can take advantage of the small changes in $f(a/W)$ during small increments in crack length. In this way, instead of calculating at each load cycle the value of $\Delta K_i = \Delta\sigma_i \cdot [\sqrt{(\pi a_i)} \cdot f(a_i/W)]$, a task that demands great computational effort, it is more efficient to hold $f(a_i/W)$ constant during a (small) percentage of crack increment $\delta a\%$, that should be specifiable by the calculation software user. The errors introduced by this procedure are non-conservative, but they decrease quickly with the value specified for $\delta a\%$. Convergence is assumed when decreasing values of $\delta a\%$ do not influence significantly the calculated crack growth. In practice, $\delta a\%$ values of the order of 0.5% result in adequate predictions for most calculations [25].

Two-Dimensional Crack Growth by the Cycle-by-Cycle Method

The 2D growth of elliptical cracks problem can be treated in manner similar to that already discussed in the ΔK_{rms} method: if in the i-th loading event the ellipsoidal crack has semi-axes a_i and c_i, under

$$\Delta K(a_i) = \Delta\sigma_i \cdot [\sqrt{(\pi a_i)} \cdot f_a(a_i/c_i,\ a_i/t,\ c_i/W)] \text{ and } \Delta K(c_i) = \Delta\sigma_i \cdot [\sqrt{(\pi a_i)} \cdot f_c(a_i/c_i,\ a_i/t,\ c_i/W)] \quad (28)$$

stress intensity ranges, and if $da/dN = F(\Delta K,\ R,\ \Delta K_{th},\ K_C,...)$ is the crack growth rule of the material, then the crack increment in this i-th 1/2 cycle is given by

$$\delta a_i = \frac{1}{2} \cdot F(\Delta K(\Delta\sigma_i, a_i, f_a), R(\Delta\sigma_i, \sigma_{max_i}), \Delta K_{th}, K_C, ...) \quad (29)$$

$$\delta c_i = \frac{1}{2} \cdot F(\Delta K(\Delta\sigma_i, a_i, f_c), R(\Delta\sigma_i, \sigma_{max_i}), \Delta K_{th}, K_C, ...) \quad (30)$$

The crack growth is calculated by the simultaneous solution of $\Sigma\delta a_i$ and $\Sigma\delta c_i$. As the crack increments δa_i and δc_i depend on both a_i and c_i, the coupled 2D growth is well characterized.

It is worth mentioning that all the comments made above about the filtering and the counting of the loading can be also applied to 2D crack growth. In the same way, the maintenance of f_a and f_c constant during a small $\delta a\%$ (or $\delta c\%$) variation in crack length is still more useful in 2D calculations, because the analytical expressions for f_a and f_c are generally very complex, and demand great numerical effort. But, besides the computational complexity, the 2D cycle-by-cycle problem does not present significant supplementary conceptual difficulties over the 1D case.

Load Cycle Interaction Effects

It is a well-known fact that interaction problems among load cycles can have a very significant effect in the prediction of fatigue crack growth. There is a vast literature proving that tensile overloads, when applied over a loading whose amplitude otherwise stays constant, can cause retardation or arrest in the crack growth, and that even compressive overloads can sometimes affect the rate of subsequent crack propagation [*33, 35, 36*].

Neglecting these effects in fatigue life calculations can completely invalidate the predictions. In fact, only after considering overload induced retardation effects can the life reached by real structural components be justified when modeling many practical problems. However, the generation of an universal algorithm to quantify these effects is particularly difficult, due to the number and to the complexity of the mechanisms involved in fatigue crack retardation, among them: plasticity-induced crack closure; blunting and/or bifurcation of the crack tip; residual stresses and/or strains; strain-hardening; crack face roughness, and oxidation of the crack faces.

Besides, depending on the case, several of these mechanisms may act concomitantly or competitively, as a function of factors such as: size of the crack; microstructure of the material; dominant stress state, and environment.

The detailed discussion of this complex phenomenology is considered beyond the scope of this work (a revision of the phenomenological problem can be found in [*33*]). Moreover, the relative importance of the several mechanisms can vary from case to case, and there is, so far, no universally accepted single equation capable of describing the whole problem. Therefore, from the designer's point of view it must necessarily be treated in the most reasonably simplified way.

A simplified model can not be unrealistic so it is worthwhile mentioning that some simplistic models are unacceptable. For example, attributing the overloads to a significant variation in the residual stress state at the crack tip in order to justify the retardation effects is not reasonable. This is mechanically impossible: the tensile yielding during the loading and the compressive yielding during the unloading close to the crack tip during fatigue crack propagation prevent any significant variation in the residual stress state at the crack tip after an overload. On the other hand, the principal characteristic of fatigue cracks is to propagate cutting a material that has already been deformed by the plastic zone that always accompanies their tips. In this way, the fatigue crack faces are embedded in an envelope of (plastic) residual strains and, consequently, they compress their faces when completely discharged, and they open alleviating in a progressive way the (compressive) load transmitted through their faces.

According to Elber [*37*], only after completely opening the crack at a load K_{op}, would the crack tip be stressed. Therefore, the bigger the K_{op}, the less would be the effective stress intensity range $\Delta K_{eff} = K_{max} - K_{op}$, and this ΔK_{eff} instead of ΔK would be the crack propagation rate controlling parameter. Most load interaction models are, although indirectly, based in this idea. This implicates in the supposition that the principal retardation mechanism is caused by plasticity induced crack closure: in these cases, the opening load should **increase** when the crack penetrates into the plastic zone inflated by the overload, **reducing** the ΔK_{eff} and stopping or delaying the crack, while the plastic zones associated with the loading are contained in the overload induced plastic zone.

It is very important to emphasize that this is by no means the only mechanism that can induce crack retardation. For example, Castro and Parks [*38*] showed that, under dominant plane strain conditions, overload induced fatigue crack retardation or arrest can occur while ΔK_{eff} **increases**. The principal retardation mechanism in those cases was bifurcation of the crack tip.

The Wheeler model is the most popular retardation model [*35*]. The model is simplistic and assumes, more or less arbitrarily, that while the loading plastic zone ZP_i is embedded in the overload plastic zone ZP_{ov}, the crack growth rate retardation depends on the distance from the border of ZP_{ov} to the tip of the crack, see Figure 11.

The retardation is maximum just after the overload, and stops when the border of ZP_i touches the border of ZP_{ov}. Therefore, if a_{ov} and a_i are the crack sizes at the instant of the overload and at the i-th cycle, and $(da/dN)_{reti}$ and $(da/dN)_i$ are the retarded and the non-retarded crack growth rate (at which the crack would be growing in the *i*-th cycle if the overload had not occurred), then, according to Wheeler

$$\left(\frac{da}{dN}\right)_{ret_i} = \left(\frac{da}{dN}\right)_i \cdot \left(\frac{ZP_i}{a_{ov} + ZP_{ov} - a_i}\right)^{\beta}, \; a_i + ZP_i < a_{ov} + ZP_{ov} \qquad (31)$$

where β is an experimentally adjustable constant. Broek [*35, 36*] mentions Wheeler's data for steels ($\beta = 1.43$) and for Ti-6AL-4V ($\beta = 3.4$), and suggests that other typical values for β are between 0 and 2.

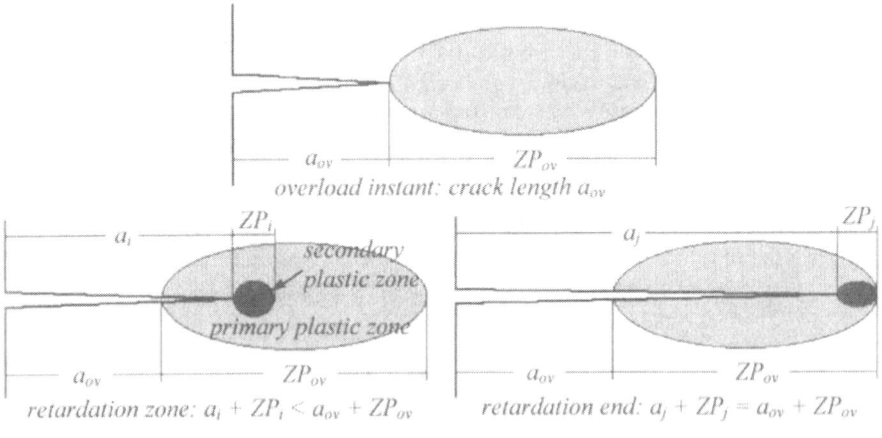

Figure 11 – *Wheeler crack growth retardation model.*

It should be noticed that this model cannot predict crack arrest that has been observed. As $ZP \approx (K_{max}/S_Y)^2$, where S_Y is the yielding strength of the material, the maximum value of the predicted retardation happens immediately after the overload, and is equal to $(K_{max}/K_{ov})^{2\beta}$, where K_{max} is the maximum load in the cycle just after the overload, and K_{ov} is the overload peak. Therefore, the phenomenology of the load cycle interaction problem is not completely reproducible by the Wheeler original model. However, also to model crack arrest, a simple modification that seems reasonable is to use a Wheeler-like parameter to multiply ΔK instead of da/dN after the overload

$$\Delta K_{ret}(a_i) = \Delta K(a_i)\cdot\left(\frac{ZP_i}{a_{ov} + ZP_{ov} - a_i}\right)^{\gamma}, \quad a_i + ZP_i < a_{ov} + ZP_{ov} \tag{32}$$

where $\Delta K_{ret}(a_i)$ and $\Delta K(a_i)$ are the values of the stress intensity factors that would be acting at a_i with and without retardation due to the overload, and γ is in general different from the original model exponent β. This simple modification can be used with any of the propagation rules that recognize ΔK_{th} to predict both the retardation and the stop of fatigue cracks after an overload (the stop occurring if $\Delta K_{ret}(a_i) < \Delta K_{th}$).

The numeric implementation of these retardation models in a cycle-by-cycle algorithm is not conceptually difficult, but it requires a considerable programming effort. To illustrate the main ideas, a simplified flow-chart of the calculation algorithm used in the **ViDa** software is shown in Figure 12.

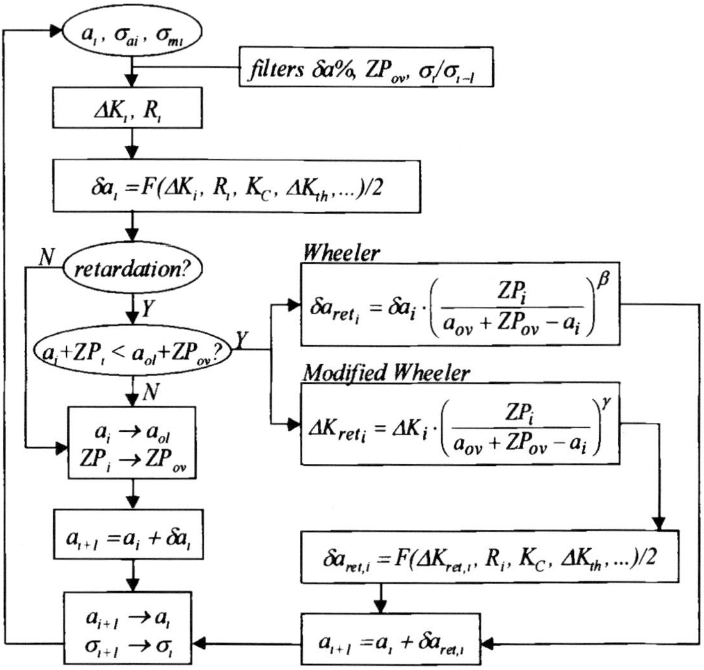

Figure 12 – *Simplified flow-chart of the calculation algorithm used in the* **ViDa** *software to predict fatigue crack propagation under complex loading.*

Some calculation details are worth mentioning. The first one refers to the use of the $\delta a\%$ filter, since crack size increments that work well otherwise can cause troubles with the retardation models, as the plastic zone sizes can be very small compared to the crack size. In order to quantify the propagation gradient inside the overload affected zone, $\delta a\%$ must be much smaller than ZP_{ov}.

A second one can save a lot of computational time when the loading is complex. Small variations in the loading amplitude do not cause experimentally detectable crack retardation, and they should not be considered as overloads in the calculation model. Therefore, a numerical filter for overloads can be profitably introduced in the algorithm, specifying that there is no overload effect if $\sigma_j/\sigma_{j-1} < \alpha$, where σ_{j-1} and σ_j are successive peaks of the loading and α is an adjustable constant (that, in the absence of better information, can be chosen as 1.25 or 1.3).

There are several other retardation models [*33, 39*], but none of those that can be implemented in a local approach code has definitive advantages over the simpler Wheeler models discussed above. This is no surprise, since single equations are too simplistic to model all the several mechanisms that can induce retardation effects. Therefore, in the same way that a curve *da/dN* vs. ΔK is experimentally measured, a propagation model can be adjusted to experimental data to calibrate the exponents of Equations 31 or 32, as recommended by Broek [*35*].

Using these same concepts, it is not particularly difficult to model retardation effects in 2D crack propagation. The idea is to maintain the fundamental hypothesis of the ellipsoidal geometry preservation, accounting for the coupled growth of the semi-axes a and c. However, as the size of the plastic zones depends on the value of ΔK, and as in general $\Delta K(a) \neq \Delta K(c)$, the retardation effects in 2D growth can be different in each direction. Thus, after each overload the value of $\Delta K_{ret}(a_i)$ and $\Delta K_{ret}(c_i)$ are calculated by

$$\Delta K_{ret}(a_i) = \Delta K(a_i) \cdot \left(\frac{ZP(a_i)}{a_{ov} + ZP(a_{ov}) - a_i} \right)^{\gamma}, \quad a_i + ZP(a_i) < a_{ov} + ZP(a_{ov}) \tag{33}$$

$$\Delta K_{ret}(c_i) = \Delta K(c_i) \cdot \left(\frac{ZP(c_i)}{c_{ov} + ZP(c_{ov}) - c_i} \right)^{\gamma}, \quad c_i + ZP(c_i) < c_{ov} + ZP(c_{ov}) \tag{34}$$

From these values, it is easy to calculate the crack growth in the two elliptic semi-axes directions during the *i*-th 1/2 cycle of the loading

$$\delta a_i = \frac{1}{2} \cdot F(\Delta K_{ret}(\Delta\sigma_i, a_i, f_a), R(\Delta\sigma_i, \sigma_{max_i}), \Delta K_{th}, K_c, \ldots) \tag{35}$$

$$\delta c_i = \frac{1}{2} \cdot F(\Delta K_{ret}(\Delta\sigma_i, a_i, f_c), R(\Delta\sigma_i, \sigma_{max_i}), \Delta K_{th}, K_c, \ldots) \tag{36}$$

Only the modified Wheeler model is presented above, but it is trivial to write similar equations for the original model. Finally, it should be mentioned that all the filtering remarks already discussed are also applied to the 2D case.

Conclusions

In this paper, a two-phase methodology was presented to predict fatigue crack propagation in generic 2D structures under complex loading. First, self-adaptive finite-elements were used to calculate, by three different methods, the fatigue crack path and the stress intensity factors along the crack length $K_I(a)$ and $K_{II}(a)$, at each propagation step. The calculated $K_I(a)$ was then used to predict the propagation fatigue life of the structure by the local approach, using the root mean square or the cycle-by-cycle integration methods, even considering overload-induced crack retardation effects in this latter one. Two complementary software programs were developed to implement this methodology. The first software package is an interactive graphics program for simulating two-dimensional fracture processes based on a finite element adaptive mesh generation strategy. The second is general purpose fatigue design software developed to predict both initiation and propagation fatigue lives under complex loading by all classical design methods. In particular, its crack propagation module accepts any stress intensity factor expressions, including the ones generated by the finite-element software. Experimental results showed that the presented methodology and its software implementation could effectively predict fatigue crack propagation in an arbitrary two-dimensional structural component.

References

[1] Shih, C.F., de Lorenzi, H.G., and German, M.D., "Crack Extension Modeling with Singular Quadratic Isoparametric Elements," *International Journal of Fracture*, Vol. 12, 1976, pp. 647-651.

[2] Rybicki, E.F. and Kanninen, M.F., "A Finite Element Calculation of Stress-Intensity Factors by a Modified Crack Closure Integral," *Engineering Fracture Mechanics*, Vol. 9, 1977, pp. 931-938.

[3] Raju, I.S., "Calculation of Strain-Energy Release Rates with Higher Order and Singular Finite Elements," *Engineering Fracture Mechanics*, Vol. 28, 1987, pp. 251-274.

[4] Bui, H.D. "Associated Path Independent J-Integrals for Separating Mixed Modes," *Journal of Mechanics & Physics Solids*, Vol. 31, 1983, pp. 439-448.

[5] Dodds, R.H. Jr. and Vargas, P.M., "Numerical Evaluation of Domain and Contour Integrals for Nonlinear Fracture Mechanics," Report UILU-ENG-88-2006, Dept. of Civil Engineering, University of Illinois, Urbana-Champaign, 1988.

[6] Banks-Sills, L. and Sherman, D., "Comparison of Method for Calculating Stress-Intensity Factors with Quarter-Point Elements," *International Journal of Fracture Mechanics*, Vol. 32, 1986, pp. 127-140.

[7] Nikishkov, G.P. and Atluri, S.N., "An Equivalent Domain Integral Method for Computing Crack-Tip Integral Parameters in Non-Elastic Thermo-Mechanical Fracture," *Engineering Fracture Mechanics*, Vol. 26, 1987, pp. 851-867.

[8] Chen, K.L. and Atluri, N. "Comparison of Different Methods of Evaluation of Weight Functions for 2D Mixed-Mode Fracture Analysis," *Engineering Fracture Mechanics*, Vol. 34, 1989, pp. 935-956.

[9] Rice, J.R., "A Path Independent Integral and the Approximate Analysis of Strain Concentration by Notches and Cracks," *Journal of Applied Mechanics*, Vol. 35, 1968, pp. 379-386.

[10] Knowles, J.K. and Sternberg, E., "On a Class of Conservation Laws in Linearized and Finite Elastostatics," *Archives for Rational Mechanics & Analysis*, Vol. 44, 1972, pp. 187-211.

[11] Atluri, S.N., "Path-Independent Integrals in Finite Elasticity and Inelasticity, with Body Forces, Inertia, and Arbitrary Crack-Face Conditions," *Engineering Fracture Mechanics*, Vol. 16, 1982, pp. 341-369.

[12] Eischen, J.W., "An Improved Method for Computing the J2 Integral," *Engineering Fracture Mechanics*, Vol. 26, 1987, pp. 691-700.

[13] Kienzler, R. and Kordisch, H., "Calculation of J1 and J2 Using the L and M Integrals," *International Journal of Fracture*, Vol. 43, 1990, pp. 213-225.

[14] Erdogan, F. and Sih, G.C., "On the Crack Extension in Plates under Plane Loading and Transverse Shear," *Journal of Basic Engineering*, Vol. 85, 1963, pp. 519-527.

[15] Hussain, M.A., Pu, S.U., and Underwood, J., "Strain Energy Release Rate for a Crack under Combined Mode I and II," *ASTM STP 560*, 1974, pp. 2-28.

[16] Sih, G.C., "Strain-Energy-Density Factor Applied to Mixed Mode Crack Problems," *International Journal of Fracture Mechanics*, Vol. 10, 1974, pp. 305-321.

[17] Bittencourt, T.N., Wawrzynek, P.A., Ingraffea, A.R., and Sousa, J.L.A., "Quasi-Automatic Simulation of Crack Propagation for 2D LEFM Problems," *Engineering Fracture Mechanics*, Vol. 55, 1996, pp. 321-334.

[18] Barsom, J.M. and Rolfe, S.T., *Fracture and Fatigue Control in Structures*, Prentice-Hall, New Jersey, 1987.

[19] Araújo, T.D.P. "Adaptive Simulation of Elastic-Plastic Fracturing," (in Portuguese), PhD dissertation, Department of Civil Engineering, Pontifical Catholic University of Rio de Janeiro (PUC-Rio), Brazil, 1999.

[20] Carvalho, C.V., Araújo, T.D.P., Cavalcante, J.B., Martha, L.F., and Bittencourt, T.N., "Automatic Fatigue Crack Propagation using a Self-adaptive Strategy," *PACAM VI - Sixth Pan-American Congress of Applied Mechanics*, Rio de Janeiro, Vol. 6, 1999, pp. 377-380.

[21] Paulino, G.H., Menezes, I.F., Cavalcante, J.B., and Martha, L.F., "A Methodology for Adaptive Finite Element Analysis: Towards an Integrated Computational Environment," *Computational Mechanics*, Vol. 23, 1999, pp. 361-388.

[22] Araújo, T.D.P., Cavalcante, J.B., Carvalho, M.T.M., Bittencourt, T.N., and Martha, L.F., "Adaptive Simulation of Fracture Processes Based on Spatial Enumeration Techniques," *International Journal the Rock Mechanics and Mining Sciences*, Vol. 34, 1997, p. 551 (abstract) and CD-ROM Paper No. 188 (full paper).

[23] Baehmann, P.L., Wittchen, S.L., Shephard, M.S., "Robust Geometrically Based, Automatic Two-Dimensional Mesh Generation," *International Journal for Numerical Methods in Engineering*, Vol. 24, 1987, pp. 1043-1087.

[24] de Floriani, L., and Puppo, E., "An On-line Algorithm for Constrained Delaunay Triangulation," *Graphical Models and Image Procesing*, Vol. 54, 1992, pp. 290-300.

[25] Meggiolaro, M.A. and Castro, J.T.P., " **ViDa 98** – a Visual Damagemeter to Automatize the Fatigue Design under Complex Loading," (in Portuguese), *Revista Brasileira de Ciências Mecânicas*, Vol. 20, 1998, pp. 666-685.

[26] Castro, J.T.P. and Meggiolaro, M.A., "Some Comments on the εN Method Automation for Fatigue Dimensioning under Complex Loading," (in Portuguese), *Revista Brasileira de Ciências Mecânicas*, Vol. 21, 1999, pp. 294-312.

[27] Hudson, C.M., "A Root-Mean-Square Approach for Predicting Fatigue Crack Growth under Random Loading," *ASTM STP 748*, 1981, pp. 41-52.

[28] Newman, J.C., "A Review and Assessment of the Stress Intensity Factors for Surface Cracks," *ASTM STP 687*, 1979, pp. 16-46.

[29] Newman, J.C. and Raju, I.S., "Stress-Intensity Factor Equations for Cracks in Three-Dimensional Finite Bodies Subjected to Tension and Bending Loads," NASA TM-85793, 1984.

[30] Anderson, T.L. *Fracture Mechanics*, CRC, 1995.

[31] Castro, J.T.P., Giassone, A., and Kenedi, P.P., "Fatigue Propagation of Superficial Cracks in Wet Welds," *Fracture, Fatigue and Life Prediction*, 1995, pp. 21-38.

[32] Castro, J.T.P., Giassoni, A., and Kenedi, P.P., "Fatigue Propagation of Semi and Quart-Elliptical Cracks in Wet Welds," (in Portuguese), *Revista Brasileira de Ciências Mecânicas*, Vol. 20, 1998, pp. 263-277.

[33] Suresh, S., *Fatigue of Materials*, Cambridge, 1998.

[34] Nelson, D.V. and Fuchs, H.O., "Predictions of Cumulative Fatigue Damage Using Condensed Load Histories," in *Fatigue Under Complex Loading*, SAE, 1977.

[35] Broek, D., *The Practical Use of Fracture Mechanics*, Kluwer 1988.

[36] Broek, D., *Elementary Engineering Fracture Mechanics*, Martinus Nijhoff 1986.

[37] Elber, W., "The Significance of Fatigue Crack Closure," *ASTM STP 486*, 1971.

[38] Castro, J.T.P. and Parks, D.M., "Decrease in Closure and Delay of Fatigue Crack Growth in Plane Strain," *Scripta Metallurgica*, Vol. 16, 1982, pp. 1443-1445.

[39] Chang, J.B and Hudson, C.M. ed., *Methods and Models for Predicting Fatigue Crack Growth Under Random Loading, ASTM STP 748*, ASTM, 1981.

P. C. McKeighan,[1] F. E. Fess II,[1] M. Petit,[1] and F. S. Campbell[1]

Quantifying the Magnitude and Effect of Loading Errors During Fatigue Crack Growth Testing Under Constant and Variable Amplitude Loading

Reference: McKeighan, P. C., Fess, F. E., II, Petit, M., and Campbell, F. S., "**Quantifying the Magnitude and Effect of Loading Errors during Fatigue Crack Growth Testing Under Constant and Variable Amplitude Loading,**" *Applications of Automation Technology in Fatigue and Fracture Testing and Analysis: Fourth Volume, ASTM STP 1411*, A. A. Braun, P. C. McKeighan, A. M. Nicolson and R. D. Lohr, Eds., American Society for Testing and Materials, West Conshohocken, PA, 2002.

Abstract: An automated data acquisition system has been developed that can be used as an independent watchdog to monitor the loads applied during a fatigue test. An algorithm to perform peak-picking based on a quadratic least squares approach has been shown to yield peak/valley magnitudes to within 1%. A damage parameter is described that assesses the impact of loading error on the resulting fatigue crack growth behavior. The peak-picking system has been validated for both constant amplitude as well as variable amplitude, spectrum loading. Overall test system performance is shown to vary significantly during the spectrum testing for the three different types of controllers: an analog controller with a computer control fatigue test system and two different types of digital controllers. This testing and analysis clearly demonstrates the importance of quantifying range error (this is not the same as end-level error), number of occurrences for each range, and overall damage content. Finally, the importance of standardizing performance metrics for spectrum testing methods is presented along with some potential recommendations for what these metrics could be and how they might be developed.

Keywords: fatigue testing, loading errors, closed loop control, peak-picking algorithm, data acquisition, variable amplitude, spectrum loading

Introduction

The latest design methodologies employ probabilistic approaches to account for any uncertainties employed in the design approach, as well as the relevant statistical distribution of these uncertainties. Material property data is often treated as a fundamental variable in these probabilistic approaches. Moreover, the variability attributed to inherent material differences may actually be an artifact of measurement error or experimental uncertainty. The work described in this paper attempts to

[1] Manager, Engineer, Research Scholar and Senior Engineering Technologist, respectively, Southwest Research Institute, 6220 Culebra Rd, San Antonio, TX, 78238-5166.

understand one of the fundamental variables involved in materials testing: the uncertainty associated with the applied loads. Fatigue crack growth (FCG) testing is highly sensitive to the applied load level since the resulting growth rate varies exponentially with load magnitude (typically to a power of three or greater). The vast majority of the work in the literature concerning variability under spectrum loading [*1-4*] has modeled the design process, not including issues associated with errors during testing.

Technology growth in the equipment used to perform materials testing in the past decade has been enormous. The modern digital servocontroller computer performs a range of control and data acquisition tasks that previously required a full rack of analog electronics, as well as at least one and possibly two, PC microcomputers. The new digital systems offer users a more streamlined and simplified approach to materials testing. However this simplification and suite of available tools can result in the user relying more heavily on the digital system's ability to perform the materials tests since (a) the analog signals from the feedback transducers are less accessible and (b) automated algorithms for continuous control loop management are available for use. As with any computer-automated task, the danger exists to lull the user into a false sense of security through this high level of automation. Finally, regardless of whether a digital or analog servocontroller system is employed, due to its inherent complexity of variable amplitude fatigue testing, the user must totally rely on the computer control system to ensure the appropriate end-levels are achieved.

A renewed interest in fatigue crack growth test variability has recently occurred with the present growth of materials testing technology. In fact, ASTM Committee E-8 on Fatigue and Fracture is currently in the midst of an FCG test round-robin, the first since the original variability study was performed 30 years ago [*5*]. The work described in this paper attempts to understand better one of the sources for this variability by providing a data acquisition system that measures maximum and minimum loads on a cycle-by-cycle basis. The system developed herein basically can act as an independent watchdog during a test, regardless of the make and model of the servocontroller or computer control system (e.g., either analog or digital). A description of this data acquisition system is provided along with a validation of the peak-picking algorithm, based upon constant amplitude sinusoidal cycling. Finally, the system is applied to the most complex type of fatigue test, crack growth under variable amplitude, spectrum loading.

Methods

Data Acquisition and Peak-Picking System

The data acquisition system that was developed is connected to an analog signal source, records the continuous, streaming data and real-time writes the peaks and valleys detected in the voltage signal. The peaks and valleys are written to an external file that can subsequently be post-processed and analyzed after testing. The system developed herein was implemented on a Pentium® microcomputer using a standard, commercial 16-bit data acquisition card (National Instruments PCI-MIO-16XE-10). The computer software system was developed in the LabVIEW (National Instruments, Version 5.1) programming environment.

The data acquisition system is configured to read in a 10 V bi-polar mode and as such has a theoretical one-bit resolution of 0.305 mV. The scan rate employed was fixed at 1000 points/second. Although faster data rates could be used, this kHz rate represents a tradeoff between the size of the data array, peak measurement accuracy, and usable frequency range (nominally up to a 25 Hz cyclic rate). A fixed data rate was employed to simplify the development and validation; the next generation system development should clearly consider a variable rate depending upon the cyclic frequency employed in the testing.

A 4000 point buffer was employed with the peak-picking algorithm. At the core of the method is the National Instruments peak detector virtual instrument (VI) available in LabVIEW Version 5.1. Although this VI and the C-code used in its implementation is considered proprietary by National Instruments, the algorithm operates on an array of data with the local peaks and valleys determined by fitting a quadratic least squares relationship to a smaller subset of the data. This subset, termed a window, is incrementally slid along the buffer data, periodically yielding the peaks and valleys. Although in theory it might have been optimal to use a sinusoidal least squares form (as opposed to quadratic), one was not available for use. In practice, this is probably less of an issue since the mechanical response of the frame filters the response in some unknown manner, thus rendering the actual waveform into some non-sinusoidal perturbation.

This process is schematically illustrated in Figure 1 with a digitized 25 Hz sinusoid. The scan window shown is 17 points and two scenarios are presented: the case where the window is symmetric to the peak as well as the asymmetric case. In practice, both of these scenarios will occur as the window incrementally moves along the data array.

The ability of this quadratic method to determine the true peak/valley magnitude is dependent upon a number of variables including sampling rate, window position, and window size. These issues will be more fully examined in a subsequent section examining the ability of the quadratic algorithm to peak detect when the underlying signal is sinusoidal.

Additional Hardware

Three different servohydraulic controllers were used, each connected to either 45 kN or 90 kN double-post load frames. One system consisted of a newer analog controller (MTS model 458) connected to a Fracture Technology Associates (FTA) fatigue and fracture test control system. The FTA system can interface with either an analog or digital servocontrol system, basically acting as an intelligent, closed loop function generator. The FTA system creates the output signal, inputs the signal from a 12-bit D/A converter to the analog controller, measures the response of the system, and subsequently adjusts the output signal accordingly. Two other commercial, digital servocontrol systems were also used during this work. The materials testing software supplied with each was used in all of the testing. These two digital systems are generically identified throughout this paper as "A" and "B" since the intent herein is to highlight performance metrics, not necessarily differences between the two.

During the experimental phase of this work, other laboratory instrumentation was also used, especially to benchmark the performance of the peak-picking system. Sinusoidal signals were created with a Hewlett-Packard 33120A function generator.

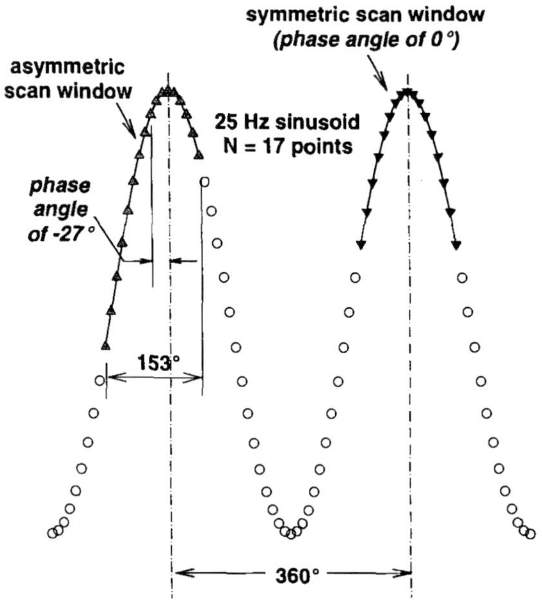

Figure 1 – *Schematic of the process used to segment the data in different scan windows.*

The signal magnitude of these sinusoidal signals was subsequently measured on a Tektronix TDS210 oscilloscope. Finally, the number of sinusoidal waveform periods output were digitally counted using a Hewlett-Packard 5300B counter. Instrumentation calibration was ensured and validated.

Experimental Test Matrix

The experiments performed are described in the series of nine trials indicated in Table 1. In each trial, the output frequency of the signal was varied from 1 to 20 Hz in discrete increments. The focus in later trials tended to be from 5 to 20 Hz, since this is the range of primary interest during materials testing. The complexity of each series increased as testing progressed: at the start only the peak-picker and a function generator were utilized as contrasted to the final series of tests using the peak-picker combined with all of the complexities associated with a materials test system. Although at times the approach is somewhat pedantic, understanding the primary contributors to error is a critical aspect of this work.

The first five trials focused on algorithm characterization under constant amplitude, sinusoidal cycling. The first two, Series A and B, employed a function generator and varied (a) peak-picking window size and (b) signal magnitude. The third series of

Table 1 – *Description of the different trials reported in this study. Each trial also included a variation of signal frequency. Trials B-I were performed with N = 19.*

Test Series	No. of Trials	Type of Loading	Signal Source	Type of Controller	Miscellaneous Comments
A	25	const. ampl.	function generator	none	vary window size (N = 13, 17, 25)
B	18	const. ampl.	function generator	none	vary signal magnitude (±0.5, 2.5, 4.5)
C	12	const. ampl.	FTA system	none	open and closed loop
D	14	const. ampl.	FTA system	analog	open and closed loop, tuned and untuned PID
E	8	const. ampl.	controller	digital A	two tuning strategies
F	9	spectrum	FTA system	none	open and closed loop
G	10	spectrum	FTA system	analog	open and closed loop
H	19	spectrum	controller	digital A	two tuning strategies
I	4	spectrum	controller	digital B	single tuning strategy

testing, identified as C, modified the signal generation, changing function generator from the electronic HP to the FTA materials test system.

During Series D and E, the influence of different servocontrollers is examined using open loop, closed loop and various tuning strategies (for the digital controllers). During this testing, whenever a servocontroller is employed, a 50% cracked compact-tension specimen (50.8 mm width) was used in the setup. Since the compact tension specimen is pin loaded, all applied loads were tensile. It should be noted that different crack sizes in the C(T) specimen would have resulted in somewhat different performance. However, this variable was omitted from further focus and deemed beyond the scope for this particular investigation.

The final series of four tests focused on performance under spectrum loading conditions. The spectrum, described in more detail later, consisted of 100 endpoints at ten distinct end-levels. Series F uses an FTA system without a servocontroller and running both open and closed feedback loop. Series G repeats this exercise but employs the analog controller in the setup. Finally, in Series H and I the two different digital controllers are used employing a range of the tuning strategies available in the software.

Results and Discussion

Theoretical Algorithm Performance

The technique of using a quadratic least squares fitting method on a portion of a digitized sinusoidal signal to determine peaks and valleys has an inherent accuracy associated with it. In order to understand the bounds of the method and its accuracy, analytical simulations of the process were performed with a FORTRAN program and least squares subroutine. For these simulations, a sinusoidal signal was created at various frequencies (2-25 Hz) and at a digital sampling frequency of 1000 points/second. Portions of the sinusoidal signal around the peak were subsequently least squares fit, varying the size of the algorithm window (5 to 30 points in increments of 5).

The results of these numerical simulations are shown in Figure 2 illustrating the peak signal prediction error as a function of where the center of the scan window is located. This scan window positioning and the details associated with measuring the phase angle are illustrated in Figure 1. A window phase angle (measured at the window center) of 0° represents a perfectly symmetric window position about the peak whereas the more negative the phase angle, the more asymmetric about the peak.

Several notable observations are apparent from the simulation data shown in Figure 2. As expected, as either cyclic frequency or the number of points in the peak-picking window increase, the error in measuring the peak magnitude also increases. Note the extremely low level of peak error (different scale) for the 2 and 5 Hz frequencies. This trend would indicate that by decreasing the number of points, the error decreases; this is not recommended, however, because a window size that is too small results in (a) multiple peak/valley counts for slower frequencies and (b) artificial peak/valleys in data that include a moderate amount of noise. The National Instruments documentation [6] provides little assistance regarding selection of the window size, providing only the maximum level that the window should be based upon the number of peaks and valleys expected to occur over a set period.

Although somewhat arbitrary, it is our judgement that a peak detector should be accurate in choosing peaks to within ±1% of peak magnitude. This estimate is based on typical component and device accuracies in modern material testing laboratories as well as a damage criteria that will be more fully described and explained later. Using this criteria, the greatest issues with the data in Figure 2 appear with the frequencies equal to or in excess of 10 Hz. For signals under 25 Hz, a window size of 15-20 points is recommended for all but the most extreme window phase angles.

Experimental Algorithm Performance

The simulations described in the preceding paragraph motivated the Series A and B trials (see Table 1) whose data is summarized in Table 2. Using a signal generator and a digital counter, 2500 cycles were fed to the peak-picker with the number of missed peaks/valleys and mV errors indicated in Table 2. Except for the highest frequency case with N = 25, all of the noted errors are less than 5 mV, 1% of the ±0.5 V signal. In fact, the vast majority of the errors indicated in Table 2 are on the order of 6- to 7-bits, arguably hardly noticeable in practice given typical noise levels in the laboratory. In accordance with the earlier simulations shown in Figure 2, the maximum error is typically observed in the highest frequency (20 Hz signal).

The methods used were effective in measuring counts to within ±1 peak/valley. This implies that only nine misses occurred over the 110 000 peaks and valleys from the constant amplitude trials listed in Table 2 (this translates to a success detection rate of 99.992%). It should be noted, however, that this performance is under excellent, noise-free conditions using a precision function generator and not subject to the noise environment of a typical servohydraulic frame. As a consequence of this data, all further peak-picking trials were performed with a window size of N = 19 points.

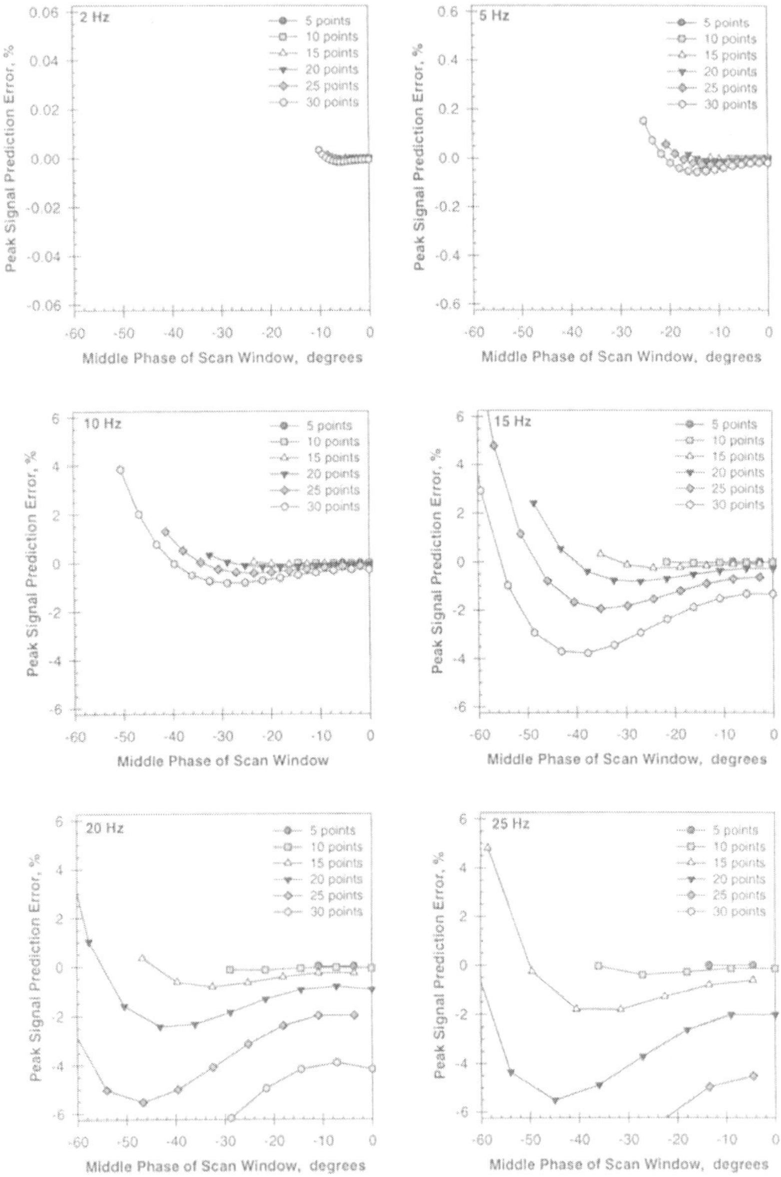

Figure 2 – *Peak detection simulation results using a quadratic polynomial algorithm for different scan window widths (points) applied to 2-25 Hz sinusoids digitized at 1 kHz. Note the error scales on the 2 and 5 Hz data sets are magnified to differentiate the results.*

Table 2 – *Peak-picking performance for Series A testing using a ±0.5 volt sinusoid generated from a precision function generator for a total of 2500 cycles for each test condition. Standard deviations on the average peak and valley magnitude were in the range of 0.5-0.9 mV.*

N, points	Freq, Hz	No. of missed peaks / valleys	Overprediction Error Peak, mV	Overprediction Error Valley, mV
13	3	0 / 1	1.97	0.64
		0 / 0	1.89	0.64
	5	3 / 1	1.88	0.61
	10	1 / 2	1.77	0.55
	15	1 / 1	1.51	0.37
	20	1 / 0	2.72	-0.18
		1 / 0	2.66	-0.19
17	2	0 / 0	1.89	0.65
		2 / 1	1.86	0.68
	3	2 / 2	1.85	0.60
		1 / 1	1.82	0.64
	5	2 / 2	1.80	0.58
		1 / 0	1.78	0.59
	10	1 / 1	1.61	0.42
	15	1 / 2	0.99	-0.16
		1 / 1	1.02	-0.20
	20	1 / 0	-0.53	-1.66
		0 / 0	-0.54	-1.65
25	2	1 / 1	1.83	0.63
		0 / -1	1.87	0.60
	20	0 / 0	-8.17	-9.28
		0 / 0	-8.17	-9.35

Similar excellent performance is indicated in Table 3 for the Series B varying signal magnitude. The theoretical percent errors shown in Figure 2 are independent of signal magnitude since the difference is a result of the inability of a second order quadratic polynomial to fit a sinusoidal response. Clearly, the overprediction error increases as the signal magnitude increases. The Series B data are plotted in Figure 3 to illustrate the peak prediction variation (on a percent basis) as a function of frequency. The similarity between the data for the three magnitudes of signal is striking, with the slightly elevated error of the ±0.5 volt signal likely a consequence of worse signal-to-noise ratio characteristics.

The experiments described above and contained in the first two series have provided excellent data suitable to validate the peak-picking algorithm implemented herein. The selection of an overall sampling rate of 1000 points/second and the use of a scan window of length 19 points appears suitable for peak detection of signals of varying magnitude and frequencies up to 20 Hz.

Table 3 – *Peak picking performance for different signal magnitudes (Series B) from a precision function generator and N = 19 points for a total of 2500 cycles for each test condition.*

Signal Magnitude, v	Freq, Hz	No. of missed peaks / valleys	Overprediction Error Peak, mV	Overprediction Error Valley, mV
±0.500	2	0 / 0	1.97	0.38
	5	1 / 2	1.90	0.41
	10	0 / 0	1.65	0.17
	15	0 / 0	0.71	-0.71
	20	2 / 1	-1.60	-2.99
±2.500	2	2 / 1	6.33	2.90
	5	1 / 1	6.08	2.64
	10	0 / 0	4.86	1.63
	15	2 / 3	0.27	-2.70
	20	-4 / -4	-11.39	-14.16
±4.500	2	-1 / 0	11.52	6.88
	5	2 / 2	11.13	6.59
	10	1 / 2	8.70	4.96
	15	0 / 0	0.42	-2.91
	20	-1 / 0	-20.22	-23.74

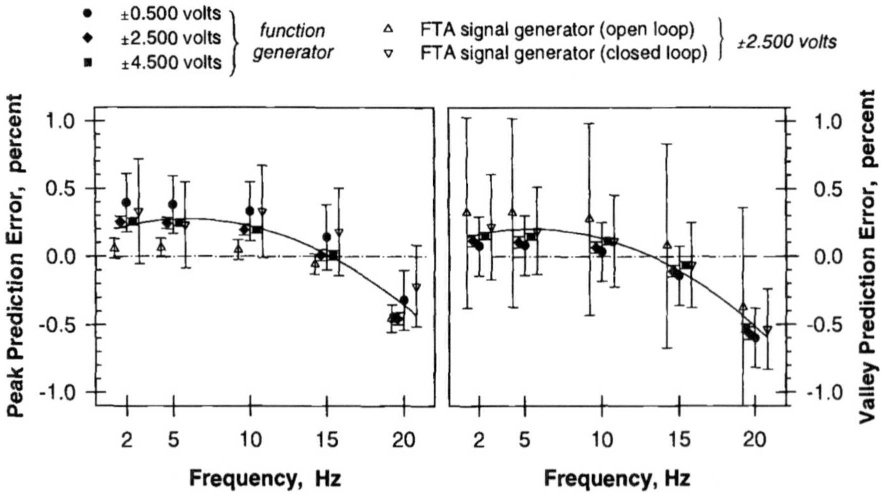

Figure 3 – *Average peak and valley detection error (percentage basis) as a function of voltage magnitude and signal generation hardware. Error bars are ±2 standard deviations for the 2500 cycle tests in test Series B and C.*

Constant Amplitude Loading

The test Series C, D, and E incorporate a materials test system function generator or digital controllers and as such had a higher level of complexity. The FTA signal generator (Series C) generally brackets the results for the precision function generator (Series B) in Figure 3. It is interesting to note that in closed loop mode, the variability of the error is less than in open loop mode for the peaks. This is not surprising, however, as in closed loop mode the system is continually correcting itself and presumably oscillating about minimum error. Although the variability may be increased, as conditions change during a materials test, it is essential to have the closed loop capability to account for such things as an increased specimen compliance. In practice, this occurs in an adaptive computer control algorithm. The narrow error bands for the open loop performance in Figure 3 are an artifact of the short time (relative to a fatigue test) and unchanging conditions represented in these trials. Moreover, it is unusual that the trend is reversed for the valleys in Figure 3; the error band decreases with closed loop control.

All of the work presented so far has not included a specimen and load transducer. Test Series D and E (Figure 4) incorporate a specimen and utilized both the FTA and a digital controller under constant amplitude sinusoidal loading. As the frequency increases, under open loop control the FTA system error rises to approximately 75 mV, or approximately 2.5% of the peak. Changing the control mode to fully closed loop results in a substantial decrease in peak error, on the order of 0.5% or less. Similar results can be observed for the digital controller A in Figure 4 using an advanced feedback (closed loop) algorithm.

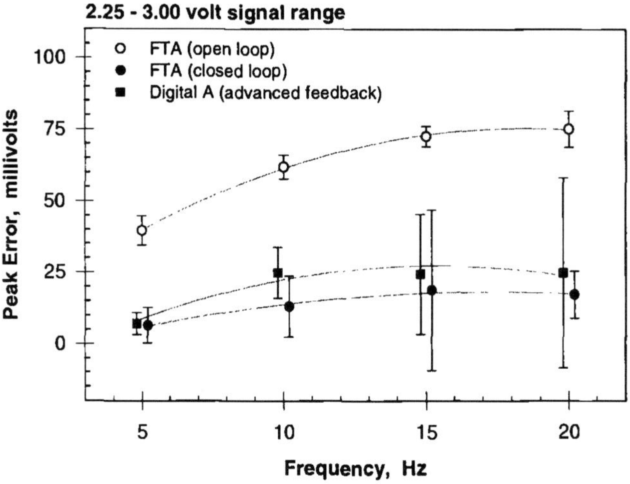

Figure 4 – *Average peak signal errors for a 3 volt signal with a specimen, analog controller/FTA system (Series D) and a digital controller system configuration (Series E). Error bars are ±2 standard deviations for the 2500 cycle test.*

Similar to the observation made earlier from Figure 3, the average error and the variance generally increase as frequency increases for the closed loop control mode. It is important to note that the end-level error in Figure 4 is approximately the same magnitude (on a percentage basis) as the peak prediction error shown in Figure 3. This similarity implies that when the materials test system is incorporated into the setup, error remains at approximately the same level (e.g., the complexity of the setup does not introduce inordinate amounts of noise to corrupt the peak-picker).

Variable Amplitude Spectrum Loading

The complexity of interpreting the peak-picking data and overall system performance increases dramatically when variable amplitude loading is employed. The essential challenges include: a) defining what a cycle is in the variable amplitude loading case and b) interpreting the implication of missed end-levels in a manner that makes sense from a materials testing viewpoint. Because both of these issues are critical for proper interpretation of the data, each will be addressed before discussing the experimental data.

Applying any type of fatigue analysis requires the amplitude of the applied cycle. As fatigue design techniques have evolved, so too have the number of available methods for interpreting cyclic data from transient end-levels [7]. A compilation of the available methods including the specific steps in each algorithm are included in ASTM Standard Practice Cycle Counting in Fatigue Analysis (E 1049). In this research work, two methods were employed: rainflow and simple range counting.

Interpreting variable amplitude errors requires understanding how errors in the applied cyclic load ΔP impact material behavior. During a fatigue crack growth test, the Paris relationship [8] provides a link between the stress intensity factor range ΔK and the fatigue crack growth rate da/dN in the following manner

$$\frac{da}{dN} = C\Delta K^{m} \tag{1}$$

where C and m, the so-called Paris coefficient and exponent, are empirically derived from a fatigue crack growth curve. For steel materials, the exponent m is typically between 3.0-3.5 [9] whereas for aluminum and titanium alloys it can be slightly higher. From fracture mechanics, we know that

$$\Delta K = \Delta\sigma\sqrt{\pi a}\, f(a/W) \propto \Delta P \tag{2}$$

which can be combined with the Paris relationship above to yield the following

$$\frac{da}{dN} \propto \Delta P^{m}. \tag{3}$$

This basically implies that fatigue crack growth testing is very sensitive to load magnitude as indicated in Figure 5 for three different Paris exponent levels. A 5% increase (or decrease) in load magnitude yields a 10-20% increase (or decrease) in

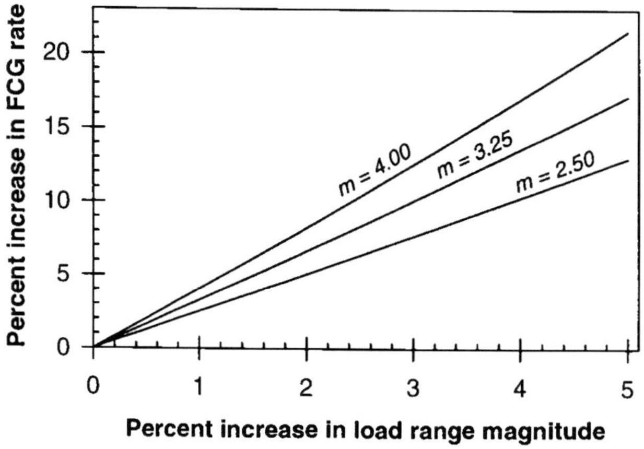

Figure 5 – *Effect of load uncertainty on the resulting fatigue crack growth rate.*

crack growth rate depending upon the actual value of the Paris exponent. Note that a 1% error in load magnitude results in less than 5% difference in crack growth rate (the source for the earlier stated 1% criteria on end-level accuracy).

Given this link between growth rate and applied load level, a parameter Γ can be defined that relates the observed "damage" based upon the actual measured ΔP to the target level in the following manner

$$\Gamma = \left[\frac{\Delta P_{actual}}{\Delta P_{target}} \right]^{3.25} \qquad (4)$$

where a conservative m = 3.25 has been assumed. A damage parameter of Γ = 0.8 indicates that over the period of focus the applied cycles resulted in 80% of the fatigue crack growth damage that would nominally have occurred had the cyclic range been exact. As should be apparent from the preceding derivation, a fundamental assumption in this damage quantification method is that the crack does not grow appreciably over the period of interest. Furthermore, the difference in fatigue crack growth rate as a function of mean stress (e.g., load ratio) is also neglected in this derivation. Although these shortcomings should be recognized, neither are considered significant in the context of a first-order, engineering estimate of load error effects during fatigue testing.

In practice, this parameter can be applied to spectra by using the magnitude of the achieved cyclic loads returned by the peak-picking algorithm. The spectra shown in Figure 6 simplify the interpretation some since, as scaled, all nine ranges are integers in the spectrum pass (it should be noted, however, that this spectrum was scaled to 2.25 volts maximum when used). As indicated previously, both rainflow and simple range counting methods were employed herein. The rainflow counting method yielded fairly

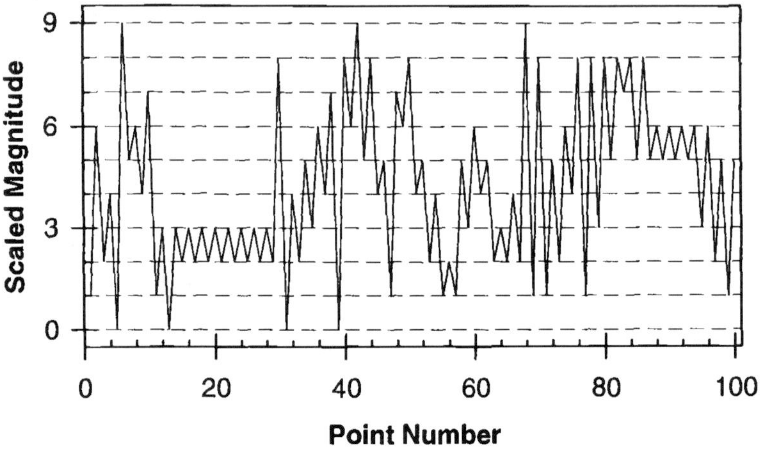

Figure 6 – *Schematic of one pass of the 100 point loading spectrum used in this work.*

inconsistent results. Some representative data in terms of a range histogram are shown in Figure 7. For the two cases considered, the damage parameter Γ is approximately 70-75%, however the end-level accuracy observed during these two tests simply did not indicate the amount of error required to lose 25-30% of the damage content. Although the definitive reason for this lack of agreement is unknown, it is suspected that noise levels created some false peaks which corrupted the very sensitive rainflow counting method. Recall that with rainflow counting, the order that the end-levels is applied is not preserved.

An alternative simple range method was subsequently applied with greater success to create a series of half-cycles. With this method, a range is defined from two sequential end-levels (the method is described more fully in ASTM E 1049). Since the spectrum was repeated, the assumption is made that these half-cycles will effectively close. However, before applying the method to the variable amplitude data, some of the earlier constant amplitude data (Series D) was processed using this approach (Table 4). Similarly, variable amplitude spectrum data are also shown in Table 4 that was produced by the FTA system fedback to itself (with no load frame or controller, Series F data). As can be clearly noted from the data in Table 4, Γ was in the range of 0.98-1.00 which is generally consistent with the excellent end-level accuracy noted with this data. This data provides confidence that the range counting method yields excellent results, at least for the situation when end-level accuracy is good.

Armed with this analysis tool, the peak-picking results from spectrum test Series G, H and I can be analyzed. Recall, these series include the highest level of complexity: variable amplitude spectrum loading, analog and digital controllers and a load frame with a specimen. The damage parameters for 50 spectrum passes and the four conditions evaluated are shown in Figure 8. The results from the analog controller and

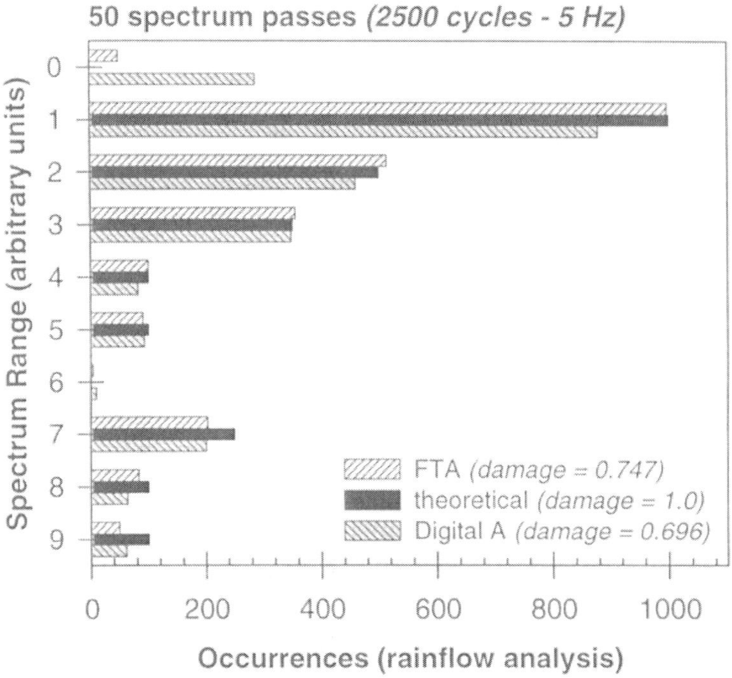

Figure 7 – *Histogram of peak-picking output for two system configurations (data is typical for results observed herein using the rainflow counting method).*

Table 4 – *Damage summation results (Series D and F) for different loading configurations using the FTA system and a simple range counting algorithm for 2500 cycles.*

Type of Loading	Configuration	*Damage Parameter Γ*			
		5 Hz	**10 Hz**	**15 Hz**	**20 Hz**
CA sinusoidal	closed loop (analog controller)	0.985	0.996	1.002	1.010
spectrum	open loop (no controller)	0.997	1.001	0.989	0.983
	closed loop (no controller)	0.994	0.990	0.989	0.976

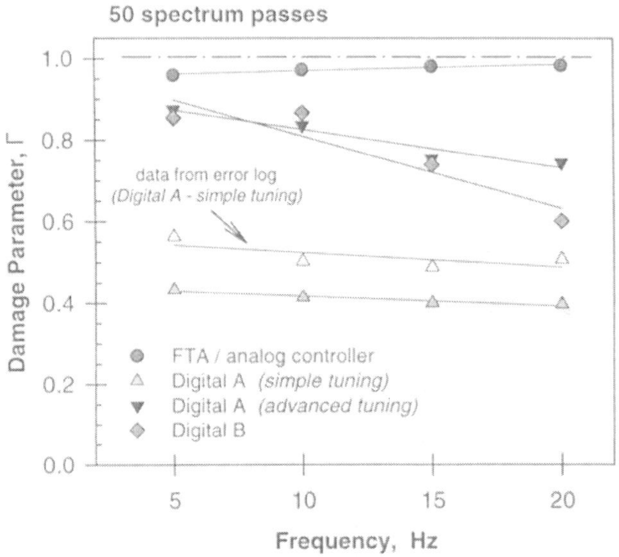

Figure 8 – *Measured system performance based upon the simple range counting algorithm and the results from the peak picking system for spectrum loading conditions (Series G, H, and I).*

the FTA computer control system are consistently excellent, typically in excess of $\Gamma = 0.95$. The FTA feedback modification appears to be effectively controlling the test as the frequency increases since the actual damage content during these trials increases.

The two digital controllers appear to provide relatively consistent data, with the exception of the simple tuning method of the "A" controller (this includes tuning at a fixed frequency, different than the test frequency) which clearly demonstrates larger error and is applying only 40% of the damage content during cycling. Moreover, the performance of the two more accurate cases for the "A" and "B" controller appears reasonable at low 5 Hz frequency (only a 15% damage content loss). However, the performance clearly degrades to a 30-40% damage content loss at the highest 20 Hz frequency. The large damage loss presumably occurs when the control algorithm can not over-drive the system sufficiently to achieve the required loads.

Because the performance of the controllers was so markedly different, an independent method for assessing overall performance **not** using the peak-picking data was undertaken. In this case, the controller-recorded end-levels, on a cycle-by-cycle basis, were processed and are shown as the unfilled points in Figure 8. The behavior of this data is consistent with that from the peak-picking algorithm, however the magnitude of the damage parameter is increased by approximately 0.1. This difference could be a consequence of using the reconstituted analog signal that is created by the digital controller to represent the recorded data.

Interpreting why such different results occurred for the different controllers requires a better understanding of the variable amplitude loading spectrum utilized in this study. A histogram of the ranges as well as an indicator of the damage content (calculated by $N\Delta P^{3.25}$ where N is the number of occurrences for the ΔP range) associated with each range is shown in Figure 9. Although the most populated ranges are the lowest four in magnitude, these ranges only account for 20% of the total damage. The vast majority of the damage, approximately 65% of the total, is contained within the highest three ranges. Careful control over these range magnitudes will ensure that the damage content applied is consistent with that expected in the variable amplitude spectrum. The data in Figure 9 clearly illustrate the multi-dimensional nature of damage under variable amplitude cycling.

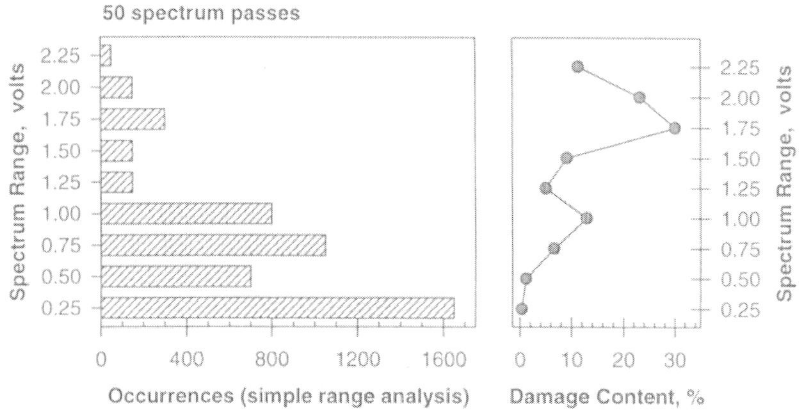

Figure 9 – *Histogram of spectrum ranges using a simple range analysis shown with the theoretical damage content of each range based on FCG behavior for the 50 passes.*

The average range error (on a percent basis) is shown in Figure 10 for three of the 15 Hz cases summarized in Figure 8. The observed range errors for the FTA and "B" digital controller are fairly tightly grouped around zero. Undoubtedly, the much better damage content of the FTA/analog controller setup observed previously is a consequence of the excellent control over the highest three ranges. Note that for the case of the "B" digital controller, all ranges **except** for the most critical, highest three are well controlled. It is interesting to note that even though the errors for the "A" digital controller are greater than observed in the "B" case, the damage parameter Γ is similar, approximately 0.75. This is a consequence of not achieving the required number of ranges as shown in Figure 11. For the case of the "B" digital controller, the number of achieved ranges is typically depleted from the theoretical. Some of these diminished ranges have apparently moved to other ranges, as indicated by the high number (700+) of ranges at the 0.0 level.

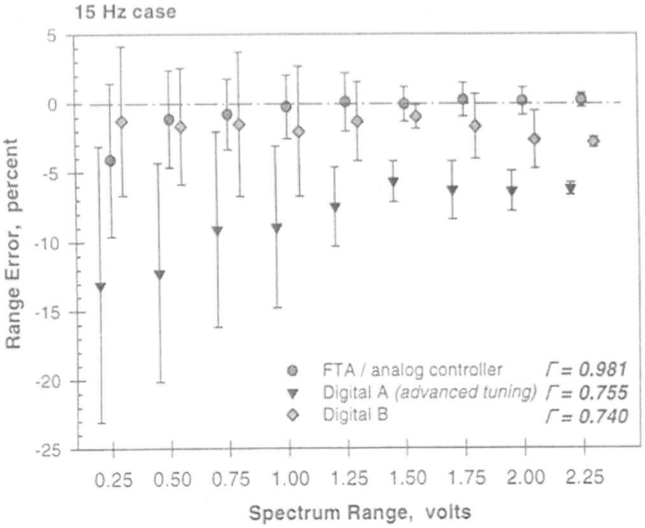

Figure 10 – *Average range percent error for three different system configurations for a 15 Hz spectrum run consisting of 50 passes. Error bars are ±2 standard deviations for each average. The data is also shown in Figure 8 at 15 Hz.*

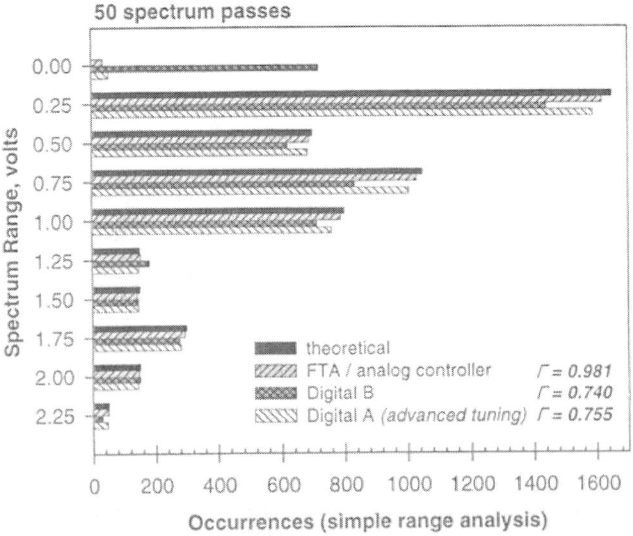

Figure 11 – *Histogram of spectrum ranges for the 15 Hz data taken from in Figure 8 (15 Hz) and shown in Figure 10. The data is derived from 50 spectrum passes.*

Summarizing Remarks

Controlling loads under variable amplitude loading conditions is clearly not a simple task. To make matters more difficult, as a technical community we have not developed the methods or the parameters to simplify this process. The data shown in Figures 8, 10, and 11 can be summarized as follows:

- in one case, the ranges and occurrences were in accordance with the desired magnitudes and the damage content approached 1.0,
- in a second case, the range magnitudes appeared fairly accurate but analysis indicated a 25% loss in damage content (due to fewer range occurrences), and
- in a third case, the range occurrences were accurate but analysis indicated a similar 25% loss in damage content (due to poor range accuracy).

Laboratories that perform spectrum testing need some method to ensure that the loads applied have been sufficiently well controlled to measure the relevant crack growth behavior. Although in theory if end-levels can be guaranteed, no additional parameter is necessary. However, this is not sufficient since some error will always occur and without quantifying the significance of that error, spectrum test results will be highly laboratory-dependent.

One recommended approach would be to ensure control by monitoring ranges, occurrences and damage content. It is believed that all three are required as a minimum to ensure that test results are independent of the methods used during the assessment. Although it might appear that focusing on end-levels would guarantee ranges, this is not necessarily the case since there still would be no metrics available to assess test performance. One possible approach to quantifying damage content has been described herein with the Γ parameter.

Determination of the applied ranges is a critical step in ensuring that the damage can be quantified (using the method described herein for damage). Moreover, as the earlier data clearly shows, an accurate range distribution is important. Criteria must yet be developed that allow one to determine how many occurrences of a particular range can be missed. It may be possible to use an approach linked to overall damage content of the range and the number of occurrences. For example, the number of occurrences that can be lost in a given range should be less than the number that can be lost to reduce the damage by 1%, for example.

The data analyzed herein should also clearly indicate to the technical community the importance of carefully examining load magnitudes when performing spectrum testing. It is critical to ensure that end-level accuracy is high; if it isn't, action should be taken either to change the test frequency or to modify the control characteristics of the system to ensure that accurate loads are achieved. It is not sufficient to simply assume that the control system is applying accurate loads. Care should also be taken to ensure that the plethora of automated tools does not create a false sense of security. Simply because an automated tuner is available on a setup does not guarantee that the tuned system will subsequently be able to apply the loads required.

Conclusions

1) A quadratic least squares fit can be used reasonably well (within 1% accuracy) as a peak detector algorithm for sinusoidal loading provided the user is careful regarding the frequency range involved, the sampling rate utilized, and the number of points used for the fit.
2) Quantifying missed end-levels during a spectrum fatigue test is not sufficient for determining the overall performance of the test. An approach focusing on the theoretical applied damage based upon fatigue and fracture mechanics principles is recommended.
3) Standardization is required in the technical community to ensure that the loading applied during a variable amplitude, spectrum fatigue test is consistent between laboratories. In particular, focusing on applied ranges, occurrences, and a parameter that quantifies the damage comparing theoretical to that applied for a given loading block should provide sufficient conditions to ensure test performance.

Acknowledgments

This work was primarily funded by the Southwest Research Institute Advisory Committee for Research (Internal Research Program R9206). Furthermore, the efforts of SwRI colleagues Messrs. J. H. Feiger, T. E. Masden, D. W. Haines, and D. L. Wagar are also acknowledged.

References

[1] Svensson, T., "Prediction Uncertainties at Variable Amplitude Fatigue," *International Journal of Fatigue*, Vol. 19, Supp. No. 1, pp. S295-S302, 1997.

[2] *Methods and Models for Predicting Fatigue Crack Growth Under Random Loading, ASTM STP 748*, J. B. Chang and C. M. Hudson, Eds., American Society for Testing and Materials, Philadelphia, PA, 1981.

[3] *Advances in Fatigue Lifetime Predictive Techniques, ASTM STP 1122,* M. R. Mitchell and R. W. Landgraf, Eds., American Society for Testing and Materials, West Conshohocken, PA, 1992.

[4] *Effect of Load Spectrum Variables on Fatigue Crack Initiation and Propagation, ASTM STP 714,* D. F. Bryan and J. M. Potter, Eds., American Society for Testing and Materials, Philadelphia, PA, 1980.

[5] Clark, Jr., W. G. and Hudak, Jr., S. J., "Variability in Fatigue Crack Growth Rate Testing," *Journal of Testing and Evaluation, JTEVA*, Vol. 3, No. 6, pp. 454-476, 1975.

[6] On-line Documentation – Peak Detector VI, National Instruments LabVIEW Version 5.1, 1999.

[7] Bannantine, J. A., Comer, J. J. and Handrock, J. L., Fundamentals of Metal Fatigue Analysis, Prentice Hall, 1990.

[8] Paris, P. C. and Erdogan, F., "A Critical Analysis of Crack Propagation Laws," *ASME, Journal of Basic Engineering*, Transactions of the ASME, Vol. D85, pp. 528-534, 1963.

[9] Hudak, Jr., S. J., Burnside, O. H. and Chan, K. S., "Analysis of Corrosion Fatigue Crack Growth in Welded Tubular Joints," *Journal of Engineering Resources Technology*, Transactions of the ASME, Vol. 107, pp. 212-219, June 1985.

David L. DuQuesnay [1]

Applications of Overload Data to Fatigue Analysis and Testing

Reference: DuQuesnay, D. L., "**Applications of Overload Data to Fatigue Analysis and Testing,**" *Application of Automation Technology in Fatigue and Fracture Testing and Analysis: Fourth Volume, ASTM STP 1411,* A. A. Braun. P. C. McKeighan, A. M. Nicolson, and R. D. Lohr, Eds., American Society for Testing and Materials, West Conshohocken, PA, 2002.

Abstract: The use of fatigue data, generated using a simple periodic overload test sequence, can provide several advantages over conventionally measured fatigue data. These include more realistic and accurate fatigue life predictions for variable amplitude loading histories and shorter testing times for prototypes and fatigue test articles provided by editing non-damaging cycles from measured loading spectra. The technique involves testing uniaxial fatigue coupons using a fully reversed periodic overload of near-yield magnitude with a fixed period in an otherwise constant amplitude sequence of high stress ratio smaller cycles. The tests can be performed in either load control or strain control with the former offering the advantage of speed and the latter the advantage of greater precision and control of the tests. The net outcome of this procedure is an effective strain life curve for a given material. When used as the basis for fatigue damage calculations, this type of data can give realistic fatigue life estimates for variable amplitude loading conditions. In addition, the endurance limit exhibited by the effective strain life data, which is referred to as the intrinsic fatigue limit, provides a material-based criterion for non-damaging cycles in fatigue. This intrinsic fatigue limit can be of large magnitude in typical engineering alloys and has a practical application as a parameter for filtering non-damaging cycles from rainflow-counted fatigue service load histories. It is shown empirically that the magnitude of the intrinsic fatigue limit can be estimated from the modulus of elasticity of the material.

This paper describes a procedure for measuring the effective strain life curve for a material and demonstrates how to apply the measured data to calculate fatigue life for variable amplitude loading spectra with the aid of computer algorithms. Finally, a technique for editing service load histories to remove non-damaging cycles from service load spectra is described.

Keywords: aluminum, cycle editing, damage, elastic modulus, fatigue (materials), fatigue limit, life prediction, non-damaging cycles, steel, variable amplitude loading

[1]Professor, Department of Mechanical Engineering, Royal Military College of Canada, PO Box 17000 Station Forces, Kingston, Ontario, Canada, K7K 7B4.

Fatigue analysis and testing have long been recognized as vital components in the development and maintenance of a wide variety of engineering structures. The transportation industries in particular have been at the forefront of advancing the state of the art in this area as land, sea and air vehicles are susceptible to fatigue failures. Although there have been advances in the understanding of the fatigue behaviour of materials, especially metals, over the past several decades, there still exists a need to verify component designs and component modifications to ensure that there exists adequate fatigue life or fatigue strength. This almost without exception involves prototype and component testing following a fatigue analysis. Such a fatigue analysis can become extremely complex; for example, in the case of a torsion bar for a vehicle suspension, complex multi-axial stress-strain states exist and the component is subjected to variable amplitude loading in service. Since the components are mass produced, there is the added complexity of variation in material properties and manufacturing processes that can both affect fatigue performance in service. Hence, components can only be proven by including a suitable number of fatigue durability tests in the design process. These tests are notoriously time consuming and are really only valid if they simulate as much of the service conditions as possible. For component testing performed under laboratory conditions there is a definite economic advantage and time savings to be achieved by keeping such tests as short as possible. One way of achieving this is to apply only the damaging load cycles from the service load spectrum to the laboratory component tests.

In order to achieve efficient design and modification processes to ensure adequate fatigue performance, it is suggested that there are two key factors. The fatigue damage analysis should be as accurate as possible and the component durability tests should be as short as possible. This paper describes how overload data, measured on simple laboratory material coupons, can be used to improve the accuracy of fatigue life prediction for components subjected to variable amplitude service loading and to edit non-damaging cycles from service load spectra to shorten testing time.

Fatigue Life Prediction

The calculation of the fatigue life of a typical *safe-life* component can be broken down into a basic five-part process, i.e., determine the (1) material properties, (2) component geometry, and (3) service load spectrum; (4) perform global load to local stress-strain transformations to produce a rainflow cycle count, followed by (5) a damage summation. For the purpose of this paper, it is assumed that the component geometry is established and the service load spectrum has been measured directly, and the local stress-strain hysteresis loops can be derived and rainflow counted precisely. Hence, the areas of concern and targets for improvement are to identify the appropriate material properties and the damage summation techniques that can be used together to obtain accurate life predictions. The concerns would be similar for a *damage tolerant* design of components; however, the material properties required would be different and crack growth calculations would replace the damage summations. This paper deals with a type of component that is relatively highly stressed and contains geometric details such as fillets and holes that induce significant stress concentrations. Such components would not normally be expected to remain in service after a crack was detected and the design is considered to be based on a crack "initiation" life determined from strain-life data rather than a crack "growth" life determined using fracture mechanics. However, it is emphasised that the fatigue process is treated as one of crack growth, and the methods described in this paper are a convenient way of analysing the early stages thereof. In this paper, the term crack "initiation" refers either to the detection of a crack in a component or the failure of a laboratory strain-life specimen.

Commercial software packages, and some in-house software packages, for fatigue life prediction generally rely on material properties determined from constant amplitude (CA) fatigue tests on laboratory coupons. Miner's linear damage rule is most often used for damage summation. The combined result of using CA data with Miner's rule is that life predictions are generally unconservative for service load spectra containing occasional large or overload cycles, as would be caused by events such as pothole strikes in an automobile suspension or the ground-air-ground excursion in an aircraft wing, etc. Such overload cycles are well known to cause subsequent smaller cycles to do more damage than they would otherwise, and to cause cycles below the constant amplitude fatigue limit to also do damage. The error in life prediction is greatest for load histories containing a large number of small cycles, most below the CA fatigue limit, and occasional overloads.

The effect of variable amplitude loading on life prediction has been studied extensively, and there are a number of viable ways of improving life prediction under these conditions. For example, the BS5400 [*1*] uses a modified stress life curve with a reduced slope in the long life region and without an endurance limit. Other modified forms of stress life and strain life curves have also been suggested [*2-4*]. Some investigators have developed non-linear damage rules that can be used with measured fatigue life data to give improved life predictions [*5,6*]. The use of overload data with a simple linear damage model [*7-9*], however, has now emerged as a convenient method for the analysis of fatigue life and it offers several advantages over other techniques. Furthermore, the method works well for a wide variety of materials, load spectra, component geometries, stress magnitudes, and mean-stress levels, etc.

Damage Model

The damage model for a *safe-life* design or life to crack detection (sometimes referred to as crack "initiation" life) is based on Miner's rule, a crack-closure derived effective strain range as the damage parameter, and overload fatigue data for the material.

The fatigue process is essentially one of short crack growth since the crack length at failure is usually less than a few millimetres and the majority of the fatigue life is spent when the cracks are of microstructural dimensions. It was hypothesized and later shown by observation, that the macro-mechanisms of crack advance and damage accumulation for short cracks were similar to those of long cracks and that crack-closure had a dominant role in the fatigue behaviour of short cracks [*10-12*]. The effective strain range, which is the proportion of the strain range for which a short crack is open, has been shown to adequately account for both mean-stress effects in fatigue and for changes in damage accumulation rates following overloads in spectrum loading applications [*7-9,12*].

Some significant differences were found between the effects traditionally observed for long cracks (damage-tolerant design) and those observed for short cracks in "safe-life" fatigue design. Overloads of all types reduced crack opening stress levels and increased damage accumulation rates for subsequent cycles. Hence, no beneficial effect (crack retardation) has been observed following a tensile (T) overload, and in fact both tensile and compressive (C) overloads of near yield magnitude had approximately the same detrimental effect on the fatigue life of laboratory coupons. Also, tension-compression (TC) and compression-tension (CT) were found to have a more severe effect on subsequent cycles than a C or T overload alone. Furthermore, the CT overload was more effective than the TC in producing damage in subsequent cycles. This is contrary to what is usually reported for long cracks where the residual plasticity left by the tensile peak of a CT overload eventually leads to higher than normal crack-closure levels that cause crack retardation [*7*]. The differences between the observed behaviour of short cracks in *safe-life* components and long cracks subjected to overloads can be explained by considering two factors: (1) The strain levels associated with the former are generally much higher than those of the latter; the observations for short cracks were made for coupons tested at relatively high strain levels, with overloads of near yield magnitude, typical of those found in the roots of geometric notches in structural components. (2) The crack length at failure in safe-life components is generally within the zone of influence (plasticity) caused by the overloads. These factors lead to relatively low levels of crack opening stress that persist for the majority of the fatigue life of the component.

Experiments show that damage accumulation rates increase immediately following an overload and remain elevated for many thousands of subsequent smaller cycles. There is a logarithmic decay in the damage per cycle following an overload with the damage per cycle approaching the steady-state value that occurs in the absence of overloads (i.e., constant amplitude loading.) However, it was also observed that for CT and TC overloads of near yield magnitude, the damage per cycle not only increased but stayed constant for about 250 cycles following the overload before beginning a slow logarithmic decay to steady-state levels [*7-9*]. There is experimental evidence that the crack opening stress level is reduced by the overload and that the variations in the effective strain range of cycles following an overload correspond to the observed changes in damage accumulation rates [*7-12*].

To use the damage model, the loading spectrum is first processed into rainflow counted stress-strain ranges occurring at the fatigue sensitive location (notch root) and the largest cycle is used to calculate the crack opening stress level, S_{op}, according to [*8*]:

$$S_{op} = \alpha S_{max}[1-(S_{max}/\sigma_o)^2] + \beta S_{min} \tag{1}$$

where S_{max} and S_{min} are the maximum and minimum stresses for the largest cycle, σ_o is the cyclic yield stress, and α and β are material dependent constants with values around 0.4 to 0.6 [*7,8,13,14*]. From this, the effective strain range is determined for all the cycles in the spectrum as illustrated in Figure 1, assuming conservatively that the crack opening stress level does not increase from this value. For service load spectra with significant amounts of crack closure build-up, as may occur when overloads are infrequent or small in magnitude, load sequence effects could make this an overly conservative assumption. It may then be essential to model the increase in crack opening stress following overloads to obtain reasonable life-prediction results for those types of load spectra. The damage for each effective strain range in the spectrum is then obtained from the material resistance curve that relates the effective strain range, Δe_{eff}, to the number of cycles to failure, N_f. A method of measuring this curve from overload tests on smooth axial coupons is described below.

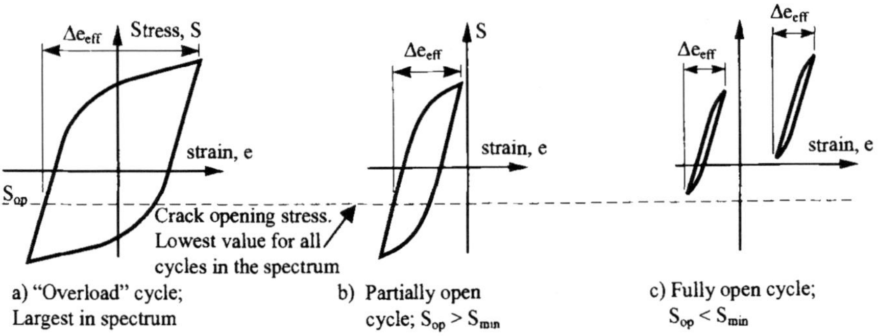

Figure 1: *Typical rainflow counted stress-strain loops for cycles in a load spectrum.*

Overload Fatigue Data – The Effective Strain Life Curve

The fatigue performance of a component will naturally be affected by the type of loading it receives in service. A severe loading spectrum will lead to shorter fatigue life than a mild loading spectrum. However, the method of calculating fatigue lives, using the model described above for example, should employ a material fatigue resistance curve that is independent of the loading spectrum.

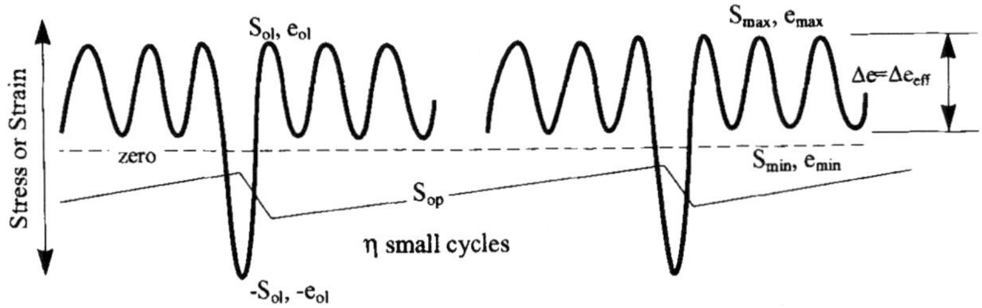

Figure 2: *Schematic of periodic overload fatigue cycles.*

A conservative approach to measuring the fatigue resistance of a material under spectrum loading is to assume that severe overloads occur with a return period short enough to ensure that the crack opening stress remains below the minimum stress of the subsequent small cycles. To keep the experimental procedure simple, it was determined empirically that a fully reversed (R=-1) CT overload of near yield magnitude, applied periodically every η cycles in an otherwise constant amplitude loading sequence of smaller cycles with the same S_{max} as the overload, would provide a reasonable estimate of the maximum damage a load cycle could do in service [7]. An example of such a test is shown schematically in Figure 2. As shown in the figure, the crack opening stress is always below the minimum stress of the small cycles and the effective strain range, Δe_{eff}, is equal to the nominal strain range, Δe, for all the small cycles in the test. The magnitude of η can be varied, but a value of about 100 has been found to give good results for steel and values of 200 to 250 have given good results for aluminum, magnesium and copper alloys. In addition, it is desirable to increase η in tests where the strain range of the small cycles is low to ensure the small cycle damage remains at least 50 percent of the total damage; otherwise the scatter in the resulting effective strain life data can become excessive.

The overload magnitude is selected such that the life of the specimens subjected to overloads alone will be about 10^4 cycles. At least three specimens should be tested in constant amplitude with the selected overload cycle to determine an average number of overload cycles to failure, N_{fo}. Assuming strain control tests are performed, subsequent specimens are tested to failure with periodic overloads and with e_{max} of the small cycles equal to e_{ol} and e_{min} varied from test to test (Figure 2) to obtain a relationship between Δe_{eff} and the number of cycles to failure, N. From the fatigue life data thus obtained, the overload damage is subtracted to determine the damage done by the small cycles from which the equivalent fatigue life of the small cycles is calculated using

$$N_f = \frac{N_s}{1 - \frac{N_o}{N_{fo}}} \tag{2}$$

where N_o is the number of overloads and N_s is the number of small cycles applied to failure ($N=N_s+N_o$). The strain range of the small cycles is varied from test to test to obtain enough data points (at least eight is recommended) to define an effective strain life curve represented by Δe_{eff} versus N_f.

Data for a 2024-T351 aluminum alloy obtained using this method from a previous study [7] and from this study are plotted in Figure 3. Also shown in the figure are constant amplitude fully reversed (R=-1) data for this alloy. It can be seen that the effective strain life curve, or "periodic overload" data, lies well below the constant amplitude curve and has a lower fatigue limit. The fatigue limit determined from the effective strain life data, Δe_i, (or $\Delta S_i = E\Delta e_i$) is considered to be an intrinsic fatigue limit for the material. Therefore, ignoring possible detrimental effects due to adverse environments which may reduce or eliminate the fatigue limit altogether, cycles below this value will be non-damaging in spectrum loading fatigue. It is noted that the two curves shown in Figure 3(a) are coincident at the overload level. Hence, the fatigue resistance of the material for cycles greater than the overload magnitude should be adequately represented by the constant amplitude strain-life data. The data in Figure 3(a) can be linearized by plotting $E(\Delta e_{eff}-\Delta e_i)$ on log-log axes as shown in Figure 3(b). In this figure, the effective strain ranges of the constant amplitude data were calculated using Equation (1) and multiplied by the elastic modulus to give units of stress for convenience.

It can be seen that the fatigue resistance of the material is well represented by an equation of the form

$$E\Delta e^{*} = E\Delta e_{eff} - E\Delta e_{i} = A(N_{f})^{b} \tag{3}$$

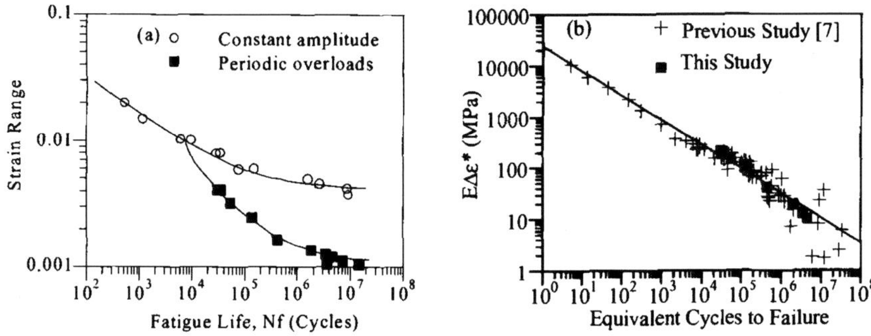

Figure 3: *Overload and constant amplitude fatigue data for 2024-T351 aluminum alloy.*

Constant amplitude and overload fatigue data measured on polished cylindrical coupons, with a gauge diameter of 8.89 mm, of 6061-T651 and 7075-T651 aluminum alloys, ZK60A magnesium alloy, and a 65/35 yellow brass are shown in Figures 4 and 5. The mechanical properties of these materials are shown later in Table 1. The overload fatigue resistance is well represented by a curve in the form of Equation (3) for each of these alloys.

The data in Figure 4(a) for 6061-T651 aluminum alloy show that the results are within normal material scatter for different magnitudes of overloads; in this case overload magnitudes of ±250 MPa and ±200 MPa, which are approximately yield strength and 80 percent of yield strength, respectively, were examined. Furthermore, as shown in the figure, tests with two sequential overloads followed by 200 high R-ratio small cycles produced data within the same scatter band as the other overload tests performed. Similar trends were found for different overload magnitudes in the 7075-T651 alloy that was tested, as shown in Figure 4(b). For the brass alloy, overload return intervals, η, varying between 50 and 200 were examined and found to give overload fatigue data in a common scatter band as shown in Figure 5(b). Hence, it is concluded that within practical limits the overload test data are not sensitive to the overload magnitudes and frequency of occurrence, provided that crack closure levels are maintained at a level below the minimum of small cycles throughout the tests.

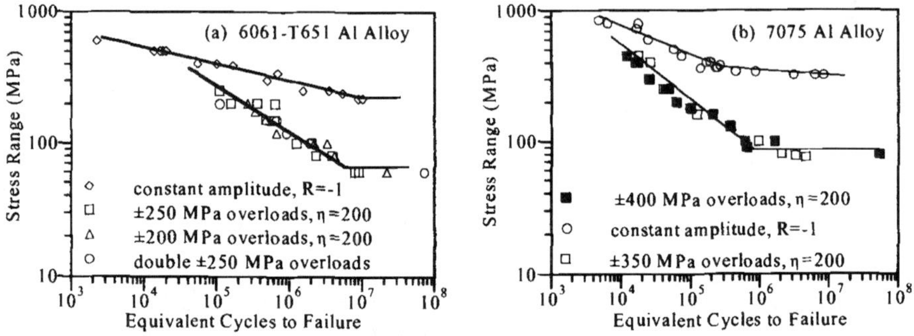

Figure 4: *Overload and constant amplitude fatigue data for aluminum alloys.*

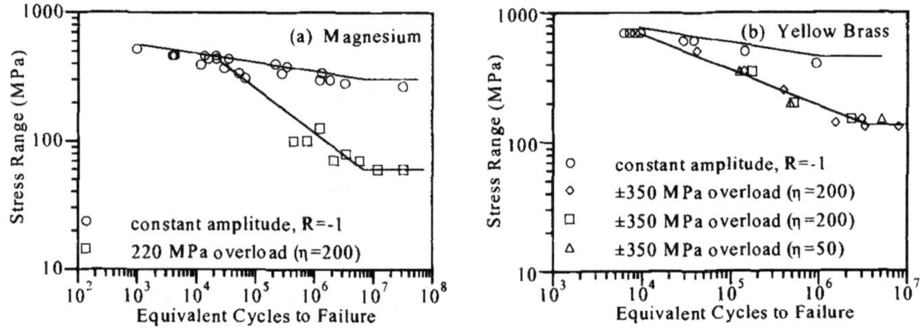

Figure 5: *Overload and constant amplitude fatigue data for other alloys.*

The overload tests can be performed in either strain control or load control as illustrated in Figure 6 for the 6061-T651 aluminum alloy. The figure shows the constant amplitude data measured in strain control at high strain levels and load control at low strain levels. The overloads used were ± 0.5% strain in strain control and ±250 MPa in load control, both with η=200. Load control offers the advantage of faster testing using higher loading frequencies while strain control allows greater precision in controlling the overload magnitudes. However, strain control may be necessary for cyclically softening materials. If the tests are performed in load control, the strains can either be measured using an axial extensometer or deduced from the cyclic stress-strain curve of the material. In a study now in progress, strain control is being used on SAE1045 steel with large plasticity overloads and the overload data are consistent with data measured in load control in previous studies.

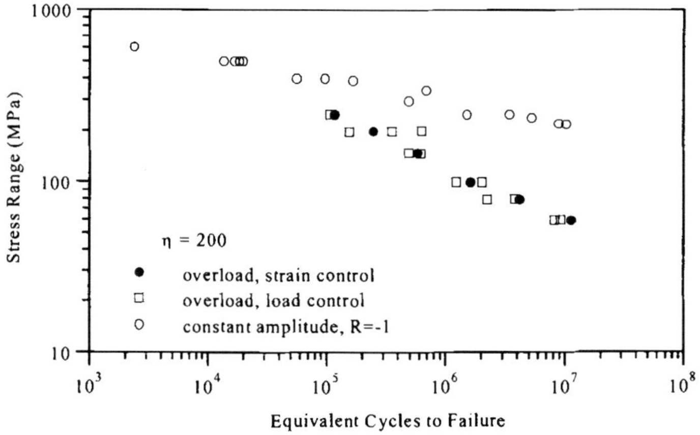

Figure 6: *Strain control and load control data for 6061-T651 aluminum alloy.*

Fatigue Life Calculations

Experimental data for 7075-T651 and 2024-T351 aluminum alloys and SAE1045 steel were collected for smooth axial coupons subjected to standard load spectra including the FALSTAFF load spectrum for fighter aircraft wings, and the SAE Log Skidder and Grapple Skiddder (bending channels) load spectra for ground vehicles. The coupons were tested in strain control with the particular load spectrum scaled linearly to give maximum strain values varying from 0.3% to about 2% to simulate notch root conditions in severe and mild notches (high and low stress concentration factors). The experimental results are plotted in Figure 7 which shows the FALSTAFF and Log Skidder spectra applied to 2024-T351 aluminum alloy and Figure 8 which shows the Log Skidder and Grapple Skidder spectra applied to SAE1045 steel.

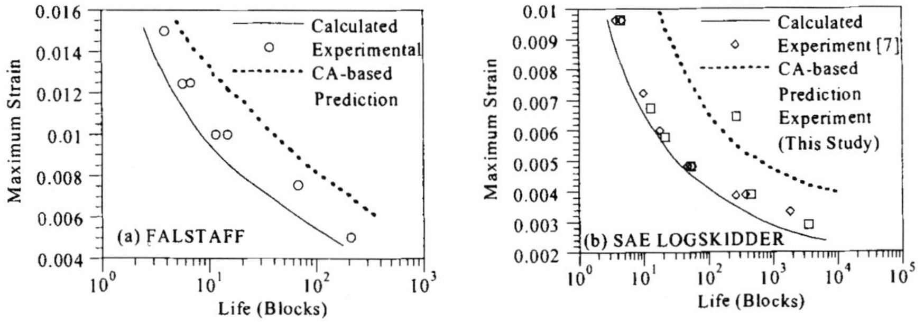

Figure 7: *Observed and calculated fatigue lives for 2024-T351 aluminum alloy subjected to a) FALSTAFF history and b) Log Skidder history.*

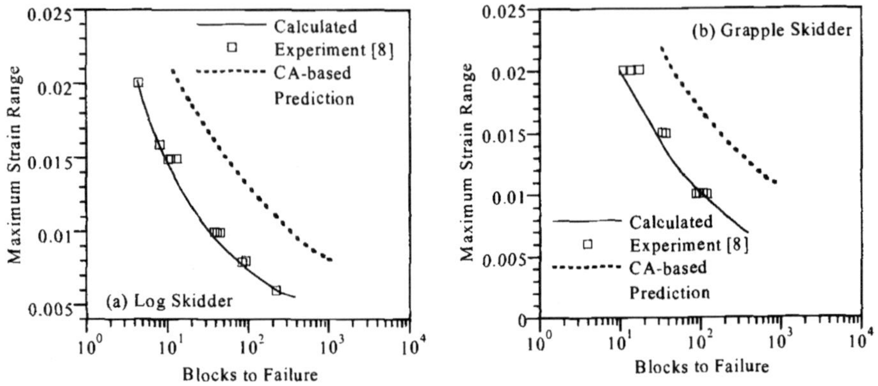

Figure 8: *Observed and calculated fatigue lives for SAE1045 steel subjected to a) Log Skidder history and b) Grapple Skidder history.*

Also shown in the figures are the calculated fatigue lives of the specimens based on overload data with the effective strain damage model described earlier in this paper for each of the materials. The calculations assume that the crack opening stress level is constant at the value set by the largest overload cycle in the rainflow counted spectrum and that damage accumulates linearly with a sum of unity at failure. In general, it is observed that the calculated lives are conservative (as expected) and provide a reasonable engineering estimate of the fatigue performance of the specimens subjected to the different load spectra examined. It has been shown that less conservative calculations are possible if the variation in crack opening stress level throughout the spectrum is modelled [*12*]. Alternatively, it has recently been shown that the crack opening stress level associated with an overload cycle with a 1/200 return frequency, rather than the largest overload cycle in the spectrum, will give less conservative and more accurate life predictions for the SAE load spectra [*13*]. For comparison with the results of the overload –based life predictions, the life predictions made using only CA data with a mean-stress correction (no overloads) are unconservative by an order of magnitude as illustrated in Figures 7 and 8.

Non-Damaging Fatigue Cycles

The intrinsic fatigue limit, Δe_i, (or $\Delta S_i = E\Delta e_i$) gives the magnitude of strain cycles that will be non-damaging in a fatigue loading spectrum. This was confirmed experimentally for the 7075-T651 aluminum alloy using the SAE Log Skidder spectrum scaled to give a maximum strain of 0.006. The spectrum, which contains 62688 reversals, was edited to remove small cycles with ranges less than given filter levels. The edited histories were then applied to smooth axial specimens under strain control to observe the fatigue life. The filter levels used, the number of reversals in the edited spectrum, and the observed fatigue lives are shown in Figure 9. The solid curve shown in the figure was fit to the data.

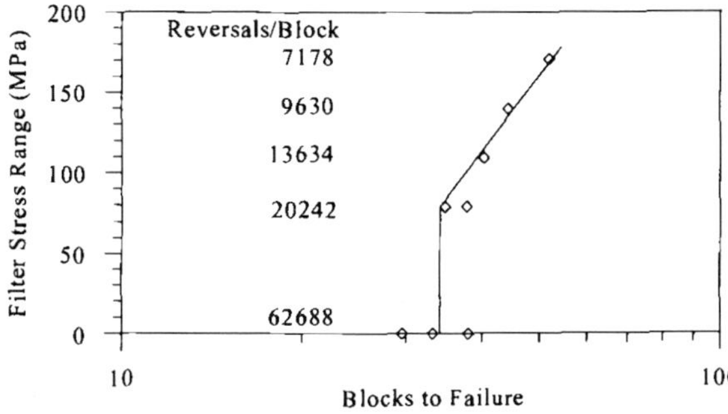

Figure 9: *Effect of small cycle omission on the fatigue life of 7075-T651 coupons subjected to the SAE Log Skidder history.*

The resulting fatigue life data indicate that cycles below the intrinsic fatigue limit, which is given by a strain range of 0.0011 and a stress range of 80 MPa for the 7075-T651, are indeed non-damaging since their removal from the spectrum did not significantly increase the fatigue life. However, as larger cycle ranges were removed from the spectrum, there was a steady increase in the observed fatigue life of the specimens, indicating clearly that the cycles removed contribute a significant amount of damage in the unedited spectrum.

Similar tests performed on smooth axial specimens of SAE 1045 steel show that the intrinsic fatigue limit of this material ($E\Delta e_i = 250$ MPa) is an appropriate criterion for filtering non-damaging cycles from the SAE spectra, as shown in Figure 10. Maximum strain ranges of 0.015 and 0.010 were considered and the fatigue lives are normalized by the average life of the unedited spectra. The number of blocks to failure for all of the data was large enough (between about 40 and several thousands) to be confident that load sequencing effects were negligible. The solid curve shown in the figure was fit to the data.

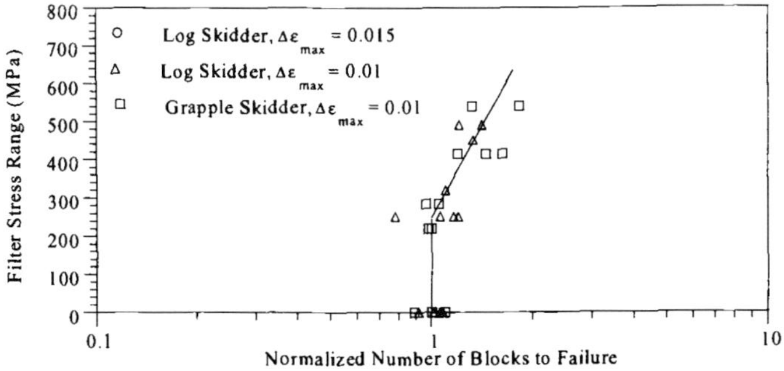

Figure 10: *Effect of small cycle omission on smooth specimens of SAE 1045 steel.*

Further experimental evidence is provided in Figure 11, which shows data for center notched plate coupons of 2024-T351 aluminum alloy subjected to filtered and unfiltered FALSTAFF and Log Skidder spectra. Again, the number of blocks to failure was large enough to be confident that load sequencing effects were negligible and the curve shown in the figure was fit to the data.

Maximum gross section stress ranges of 337 MPa and 487 MPa were considered. The components had a stress concentration factor, K_t, of about 3 and this material has an intrinsic fatigue limit stress range of 80 MPa. The crack initiation life of the notched coupons was observed using plastic replicas taken periodically to determine the number of cycles needed to form a visible surface crack of about 1 mm emanating from the notch root. Based on the K_t of the coupons, the expected stress range of non-damaging cycles should be $\Delta S_i/K_t$ or roughly 30 MPa for these test conditions. The data of Figure 11 show that this does provide a conservative level for editing non-damaging cycles from the load spectra.

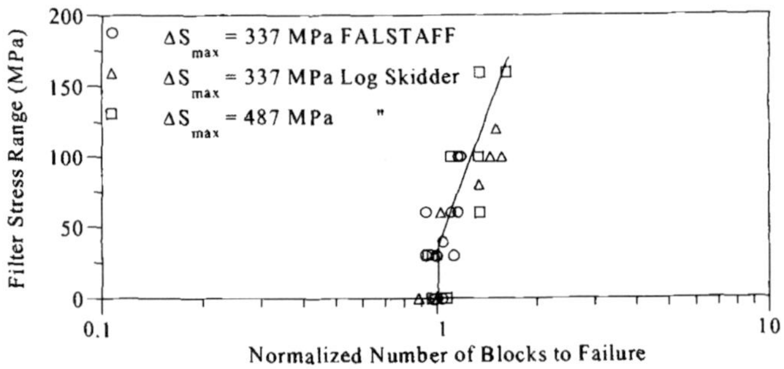

Figure 11: *Effect of small cycle omission on notched specimens of 2024-T351 aluminum alloy.*

Table 1: *Material Properties for Various Alloys*

	Normalized 1045 Steel	CSA G40.21 Steel	CSA G40.16M Steel	2024-T351 Al	6061-T651 Al	7075-T651 Al	7050-T74511 Al	6082-T651 Al	65/35 Brass	ZK60A-T5 Magnesium
Source	[*8*]	[*14*]	[*15*]	[*7*]	Fig. 5	Fig. 5	[*16*]	[*7*]	Fig 5.	Fig 5
Elastic Modulus (MPa)	205000	206000	207000	72400	68200	72000	71000	66140	98000	45000
Ultimate Tensile Strength (MPa)	745	548	650	510	290	580	520	300	545	367
Yield Strength (MPa)	460	372	450	380	240	510	450	250	400	320
Constant Amplitude R=-1 Fatigue Limit (MPa)	±300	±265	±185	±125	±110	±175	±190	±120	±200	±35
Intrinsic Fatigue Limit Stress Range (MPa)	250	275	260	80	70	85	75	90	140	60

Estimate of the Intrinsic Fatigue Limit

It was noted empirically that the magnitude of the intrinsic fatigue limit stress range (E Δe_i) was approximately constant at 80 MPa for four aluminum alloys (2024, 6061 and 6082, and 7075) with a wide range of mechanical properties. Furthermore, it was found that a variety of steels, including a CSA G40.21 structural steel [*14*], a CSA G40.16 Gr. 400M steel [*15*] and normalized SAE1045 steel [*8*] all had intrinsic fatigue limits of about 250 MPa. Since the ratio of the elastic moduli of these materials is approximately the same as the ratio of their intrinsic fatigue limits it was speculated that the intrinsic fatigue limit was related to the elastic modulus. A similar relationship was shown to exist between the intrinsic threshold stress intensity factor range, ΔK_i, and the modulus of elasticity of several metals [*17*]. There is a similarity between the intrinsic threshold stress intensity factor and the intrinsic fatigue limit stress range because both parameters represent the condition for which a "crack closure-free" fatigue crack will just grow. In the present study, an empirical analysis of intrinsic fatigue limit data for a variety of alloys, including several aluminum alloys, a magnesium alloy, a copper alloy, and several steels, showed a linear relationship between the modulus of elasticity, E, and the intrinsic fatigue limit stress range, E Δe_i, as shown in Figure 12. Other mechanical properties, such as yield stress, tensile strength, and constant amplitude endurance limit, which are listed in Table 1, did not correlate well with the intrinsic fatigue limit. Therefore, it is suggested that in the absence of experimental data, the magnitude of the intrinsic fatigue limit stress range, and hence a criterion for editing non-damaging cycles from service load spectra, can be estimated from the modulus of elasticity of the material. Alternatively, this also suggests that the intrinsic fatigue limit strain range, Δe_i, is itself constant.

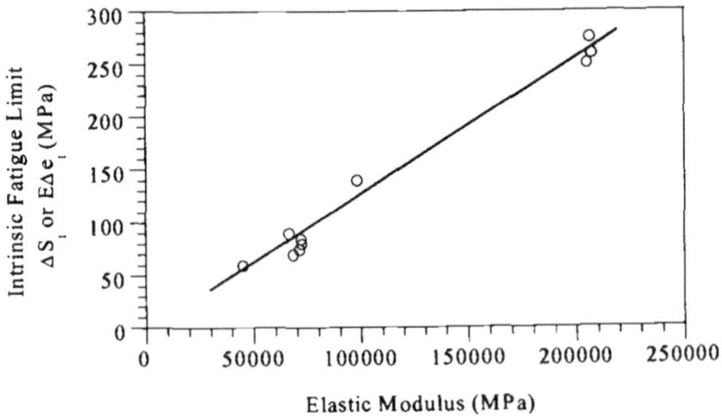

Figure 12: *Relationship between intrinsic fatigue limit and elastic modulus.*

Conclusions

1. The fatigue resistance of aluminum alloys under spectrum loading can be measured using a relatively simple periodic overload test sequence applied to material coupons. The tests can be performed in either load control or strain control. The results are not sensitive to the overload magnitude or frequency of occurrence provided that the overloads are sufficiently large and frequent to prevent crack closure build-up in subsequent cycles. Fully reversed overloads of near yield magnitude occurring every 200 cycles were used successfully for a variety of metals.
2. The overload data can be used with an appropriate damage parameter to calculate fatigue life of components subjected to spectrum loading in service.
3. The fatigue limit measured in the overload tests is an "intrinsic" fatigue limit that represents the magnitude of small cycles that are non-damaging in spectrum loads. In the absence of experimental data, the magnitude of the intrinsic fatigue limit can be estimated from the elastic modulus of the material.

Acknowledgments

Financial support by the Department of National Defence, Government of Canada is gratefully acknowledged. The author is grateful to the many Officer Cadets at the college who helped in collecting the experimental data presented.

References

[1] British Standards Institution, BS 5400, "Steel Concrete and Composite Bridges," Part 10, *Code of Practice for Fatigue*, 1980.

[2] Kikukawa, M., Jono, M. and Murata, Y., "Effect of Mean Stress and Cumulative Damage Under Service Loadings Including Stresses Below the Fatigue Limit," *Materials, Experimentation and Design in Fatigue, Proceedings of Fatigue '81*, Westbury House, IPC Business Press Ltd., 1981, pp. 308-317.

[3] Yan, X., Cordes, T.S., Vogel, J.H. and Dindinger, P.M., "A Property Fitting Approach for Improved Estimates of Small Cycle Fatigue Damage," SAE Paper # 920665, Society of Automotive Engineers, Philadelphia, PA 1992.

[4] Conle, A. and T.H. Topper, "Overstrain Effects During Variable Amplitude Service History Testing," *International Journal of Fatigue*, Vol.2, No.3, 1980, pp. 130-136.

[5] Plumtree, A. and Shen, G., "Cyclic Deformation and Life Prediction Using Damage Mechanics," *Constitutive Laws of Plastic Deformation and Fracture*, A.S. Krause *et al.*, Eds, Kluwer Academic Publishers, The Netherlands, 1990, pp. 77-85.

[6] Golos, K. and Ellyin, F., "Generalization of Cumulative Damage Criterion to Multilevel Cyclic Loading," *Theoretical and Applied Fracture Mechanics*, Vol.7, 1987, pp.169-176.

[7] DuQuesnay, D.L., Topper, T.H., and Jurcevic, R., "A New Fatigue Life Prediction Model for Variable Amplitude Loading," Localized Damage II, M.H. Aliabadi *et al.*, Eds., *Computational Mechanics Publications*, Southampton, U.K., Vol.1, Fatigue and Fracture Mechanics, pp.103-120.

[8] DuQuesnay, D.L., Pompetzki, M.A. and Topper, T.H., "Fatigue Life Predictions for Variable Amplitude Strain Histories," *SAE Transactions*, Vol. 102, No. 5, 1993, pp.455-465.

[9] DuQuesnay, D.L., MacDougall, C., Dabayeh, A. and Topper, T.H., "Notch Fatigue Behaviour as Influenced by Periodic Overloads," *International Journal of Fatigue*, Vol.17, No.2, 1995, pp.91-99.

[10] McClung, R.C. and Sehitoglu, H., "Closure Behaviour of Small Fatigue Cracks Under High Strain Fatigue Histories," ASTM STP 982, Newman, J. and Elber, W. Eds., American Society for Testing and Materials, West Conshohocken, PA, 1988, pp.279-299.

[11] Vormwald, M. and Seeger, T., "The Consequences of Short Crack Closure on Fatigue Crack Growth Under Variable Amplitude Loading," *Fatigue and Fracture of Engineering Materials and Structures*, Vol.14, No.2/3, 1991, pp.205-225.

[12] Dabayeh, A., and Topper, T.H., "Changes in Crack-Opening Stress after Underloads and Overloads in 2024-T351 Aluminum Alloy," *International Journal of Fatigue*, Vol.17, No.4, 1995, pp.261-269.

[13] Kahlil, M., DuQuesnay, D.L. and Topper, T.H., "Prediction of Crack-Opening Stress Levels for Service Loading Spectra," *Application of Automation Technology in Fatigue and Fracture Testing and Analysis: Fourth Volume, ASTM STP 1411, A. A. Braun. P. C. McKeighan, A. M. Nicolson, and R. D. Lohr, Eds.*, American Society for Testing and Materials, West Conshohocken, PA, 2002.

[14] DuQuesnay, D.L., "Fatigue Damage Accumulation in Metals Subjected to High Mean Stress and Overload Cycles," Ph.D. Thesis, Dept. of Civil Engineering, University of Waterloo, Waterloo, Ontario, 1991.

[15] DuQuesnay, D.L., Heffernan, P.J. and Howard, A.B., "Cyclic and Fatigue Behaviour of a Micro-Alloyed Reinforcing Steel," *Proceedings of the 15th Canadian Congress of Applied Mechanics*, B. Tabarrok and S. Dost, Eds., Vol.1, 1995, pp.194-195.

[16] Lynn, A. and DuQuesnay, D.L., "Computer Simulation of Variable Amplitude Fatigue Crack Initiation Behaviour Using a New Strain Based Cumulative Damage Model," submitted to *International Journal of Fatigue*, 2001.

[17] Petit, J., de Fouquet, J. and Henaff, G., "Influence of Ambient Atmosphere on Fatigue Crack Growth Behaviour of Metals," Laboratoire de mécanique et de physique des matériaux, URA 863 CNRS ENSMA, Site du futuroscope, B.P.109, Chasseneuil du Poitou, 86960 Futuroscope Cedex.

Nicolas Gérard,[1] Narayanaswami Ranganathan,[1] René Leroy,[1] Mohamed Mazari,[2] and Bel-Abbes Bachir-Bouiadjira[2]

Fatigue Crack Initiation Life Estimation at a Notch: A New Software

---

Reference : Gérard, N., Ranganathan, N., Leroy, R., Mazari, M., and Bachir-Bouiadjira, B-A., **"Fatigue Crack Initiation Life Estimation at a Notch: A New Software,"** *Applications of Automation Technology in Fatigue and Fracture Testing and Analysis: Fourth Volume, ASTM STP 1411,* A. A. Braun, P. C. McKeighan, A. M. Nicolson, and R. D. Lohr, Eds., American Society for Testing and Materials, West Conshohocken, PA, 2002.

Abstract: A New software for the estimation of the fatigue crack initiation life in notched components has been developed, based on the local strain approach. This new implementation includes a 'short crack model'. The crack driving force is evaluated by means of the stress intensity factor range for short cracks emanating from the notch, and by using an intrinsic crack growth law. A graphical user interface (GUI) has been implemented, that facilitates identification of coefficients for the strain-life curve and the stress-strain curve. The aim of this paper is to present a progress report about this software, indicating the new features and possible future improvements.

Keywords: notch, crack initiation, fatigue life, local strain, short crack, prediction, fracture mechanics.

Nomenclature

K_T	Stress concentration factor
$\Delta\sigma$	Local stress range (at notch-tip)
$\Delta\varepsilon$	Local strain range
$\Delta\sigma_\infty$	Remote tensile stress range
ΔK_I	Stress intensity factor range (mode I crack growth)
Q_K	Crack shape factor
f	Crack length correction factor

---

[1]Research Scholar, Professor, and Associate Professor, respectively, Laboratoire de Mécanique et Rhéologie, Université François Rabelais de Tours, Ecole d'Ingénieurs de Tours, 7 avenue Marcel Dassault, B.P. 407, 37204 TOURS Cedex 3 (France).
[2]PhD Students and Associate Professors, Laboratoire de Mécanique et Matériaux, IGM, Université DL de Sidi-Bel-Abbes, B.P. 89, 22000 SIDI-BEL-ABBES (Algeria).

ε'_f	Fatigue – ductility coefficient
c	Fatigue – ductility exponent
σ'_f	Fatigue strength coefficient
b	Basquin's exponent
K'	Cyclic strength hardening coefficient
n'	Cyclic strength hardening exponent
E	Young's modulus
U	Crack opening ratio

Introduction

Historically, crack initiation life at notches has been widely studied, since the need exists for precise dimensioning and life predictions enhanced with the necessary lightening of structures. The prevailing trend is to evaluate fatigue life of structures, including crack initiation and propagation phases, thus including advantages of both methods, without any of their drawbacks [*1*]. This evolution was made possible by the introduction of techniques such as strain-life approaches, which are more efficient for initiation life predictions than stress-life ones.

The limits of the latter were however extended by models that involve improvements of the K_f factor based stress-life curves [*2*]. Such approaches lead only to the prediction of the fatigue limit of notched components. The material is described using q, the notch sensitivity factor, a material-dependent parameter. Its empiricism limits its application, and in particular, such a parameter enables only the determination of whole fatigue life reduction and requires the knowledge of the stress-life curve of the material.

The local strain approach has the advantage that it enables quite easy predictions of the crack initiation life at notches. Its drawbacks can be attributed to the underlying principle, i.e., a crack is initiated at a notch when a fictitiously smooth specimen (of same material) is broken when submitted to the same strain history as the material located at the notch tip. Whereas it enables a reasonably accurate prediction of the macro-crack initiation in most cases, no information is available about the crack length.

This paper presents a new model, based on the assumption that the crack initiation life at a notch can be associated to the propagation life of a short crack, from a typical microstructural feature size up to a detectable size. Results obtained with this approach (incorporated in the presented software) will be compared with those obtained by the conventional strain life approach, and respective advantages and drawbacks will be exposed.

The Local Strain Approach and the Short Crack Model: General Presentation

The Local Strain Approach

One part of the computer program is common between the local strain approach computation (developed by G. Glinka [*3*]) and the short crack model: it involves load spectrum processing (Rain-Flow cycle counting method) and determination of the local strain history, i.e., the strain history at the notch-tip. Since the Rain-Flow cycle counting method is one of the most popular methods encountered in the literature, details will not be given here, but the reader can refer to the corresponding paper [*4*].

Regarding the strain history computation, both Neuber and Glinka methods are available. The former is based on the following equality, i.e., Neuber's equation at a notch,

$$K_T^{\,2} = K_\sigma . K_\varepsilon \tag{1}$$

where

K_ε= local strain concentration factor and
K_σ= local stress concentration factor.

Neuber's equation, coupled with a cyclic stress-strain response curve model (Ramberg-Osgood constitutive Equation (2) in our case) lead to a system of two equations, linking $\Delta\sigma$, the local stress range and $\Delta\varepsilon$ the local strain range for each cycle. Its numerical resolution gives thus the knowledge of the local strain and stress histories.

$$\frac{\Delta\varepsilon}{2} = \frac{\Delta\sigma}{2E} + \left(\frac{\Delta\sigma}{K'}\right)^{\frac{1}{n'}} \tag{2}$$

The Glinka energy method is based on a quite similar principle, and a system of two equations is obtained, relating $\Delta\sigma$ and $\Delta\varepsilon$. One of these equations remains the Ramberg-Osgood Equation (2), whereas the second one is given by the assumption that the elastic-plastic energy released at the notch-tip, W_p equals the elastic energy released in a fictitious elastic case, W_e. This assumption is geometrically symbolized in Figure 1.

This assumption leads to the second Equation (3). A comparison of both ways of evaluating local stress and strain histories was done by Glinka [*5*], and they give quite close results, comparable with finite element method (FEM) estimation.

$$\frac{(K_T . \Delta\sigma_\infty)^2}{2E} = \frac{\Delta\sigma^2}{4E} + \frac{\Delta\sigma}{n'+1}\left(\frac{\Delta\sigma}{2K'}\right)^{\frac{1}{n'}} \tag{3}$$

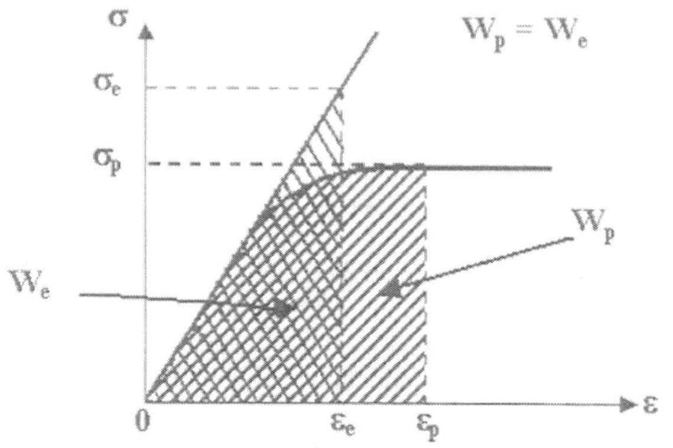

Figure 1 – *Glinka's energy-based model.*

The local strain approach uses these calculation methods to evaluate the local strain history, and a simple linear damage summation law is assumed. The damage due to each elementary cycle is taken from normalized tests on smooth specimens submitted to fully reversed strain amplitude (R_ε=-1). It is represented by an elementary damage variable $D_i = \frac{1}{N_i}$ (N_i being the failure life obtained for a smooth specimen submitted to the same strain amplitude). The crack is assumed to be initiated when the sum of each elementary damage $\sum_{i=1}^{N} D_i = D$ equals 1.

In the software, the following mean stress corrections are available: Morrow, Manson-Halford and Smith-Watson-Topper. The Manson-Coffin expression for the strain-life curve is available too. Details of these corrections are given in [*6*]. These corrections are applied only for strain amplitudes lower than the elastic-plastic transition limit.

The Short Crack Model

A new computer program was implemented to enable the crack initiation life estimation at a notch using a 'short crack model'. Crack initiation life is defined as the crack growth period up to a macro crack size (optically detectable). The crack driving force evaluation is based on the fracture mechanics parameter ΔK, related to the crack advance by an intrinsic crack growth law.

Although certain authors advocate not to use the stress intensity factor range as crack driving force for short cracks (because of linear elastic fracture mechanics –LEFM-breakdown), expressions of stress intensity factor dedicated to short cracks at notches remains a common subject of publication. Among these works, one can find those of Newman and Raju [*7*], Lukas [*8*], and Kujawski [*9*]. The former remains a reference for the concerned geometry, and the latter is a simple geometrical expression of it, using an improvement suggested by [8]. Tested on several other geometric configurations, Kujawski's expression gave quite accurate results compared with FEM calculations.

This expression was thus chosen to be implemented in the computer program, in such a manner that the elastic term [$K_T . \Delta\sigma_\infty$] in Equation (4) was replaced by the local stress amplitude evaluated by Glinka's method using elastic-plastic behavior of material.

$$K_I = Q_K . f . \frac{K_T . \Delta\sigma_\infty}{2}\left[\left(1+\frac{a}{\rho}\right)^{\frac{-1}{2}} + \left(1+\frac{a}{\rho}\right)^{\frac{-3}{2}}\right] . \sqrt{\pi a} \quad (4)$$

$$f = \begin{cases} = \left(1+\left(\dfrac{\tan\dfrac{\pi}{2.K_T}}{2.8}\right)\right)\left(\dfrac{a}{\rho} - 0.2\right) \text{for } a \succ 0.2.\rho \\ = 1 \text{ for a} \leq 0.2.\rho \end{cases} \quad (5)$$

where

K_I = stress intensity factor for short crack (Mode I opening),
a = crack length,
ρ = notch-tip radius,
Q_K = geometry factor (0.65 for a semi-circular crack, 1.22 for a through-thickness crack),
f = correction factor depending of the crack length to notch-tip radius ratio, and
$\Delta\sigma_\infty$ = remote tensile stress amplitude.

The underlying idea is to evaluate the crack propagation life (up to a detectable crack size, defined as 'initiation size') of the studied notch geometry using the crack driving force ΔK. The initial crack size is the typical microstructural feature size, such as an inclusion or a cavity. In fact, in the short crack model unlike the local strain approach, there is no 'initiation phase' corresponding to damaging of the material at the notch-tip that leads to the initiation of a macro-crack. On the contrary, a crack located at an existing flaw is assumed to propagate up to a detectable size.

Once the crack driving force parameter is evaluated, the crack advance per cycle has to be computed. The crack growth law used is an 'intrinsic law' correlating the crack growth rate to the effective stress intensity factor range for the studied material. The main hypothesis underlying this approach is based on the assumption that short cracks propagate without any crack closure. The curve that is used relates thus the effective stress intensity range (for long cracks) to the crack advance per cycle.

Nevertheless, inclusion of crack closure effect is possible, and at constant amplitude loading for Al 7075-T6 (at negative stress ratio), the evolution of U (defined in Eq. 6) as a function of ΔK_{eff} after Chong-Myong Pang [10] was tested, and results will be presented later in this paper.

$$U = \frac{\Delta K_{eff}}{\Delta K} = \frac{K_{max} - K_{op}}{K_{max} - K_{min}} \tag{6}$$

with

K_{max} = maximum stress intensity factor,
K_{min} = minimum stress intensity factor, and
K_{op} = opening stress intensity factor.

The flow chart of the algorithm developed is given in Figure 2.

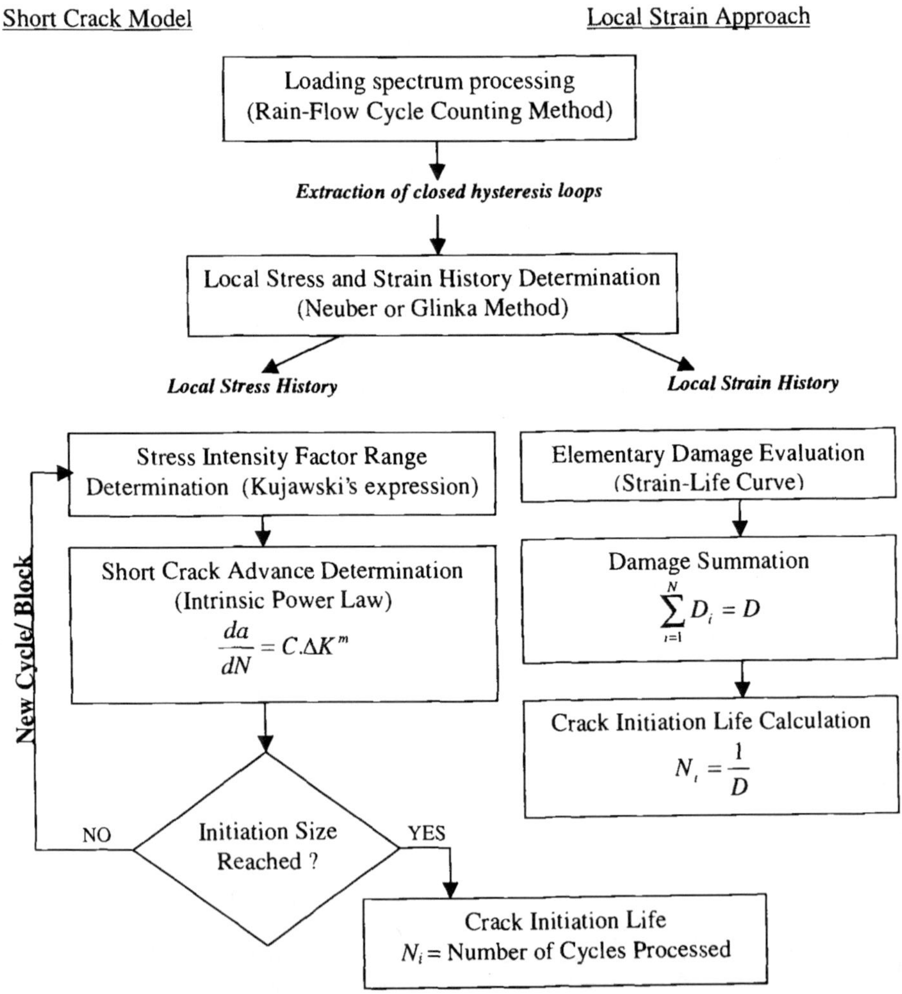

Figure 2 - *Flow-chart of the model/software.*

Prediction of the Experimental Scatter

Another particularity of this new software is the possibility of evaluating scatter in predicted lives associated with the scatter in the material properties. It was found [*6, 11*] that the prediction of experimental scatter is possible, provided input parameters are allowed to vary within their confidence range. An example is shown below (Figure 3), which permits the identification of three strain-life relationships using mean, minimum and maximum lives at a given stress level.

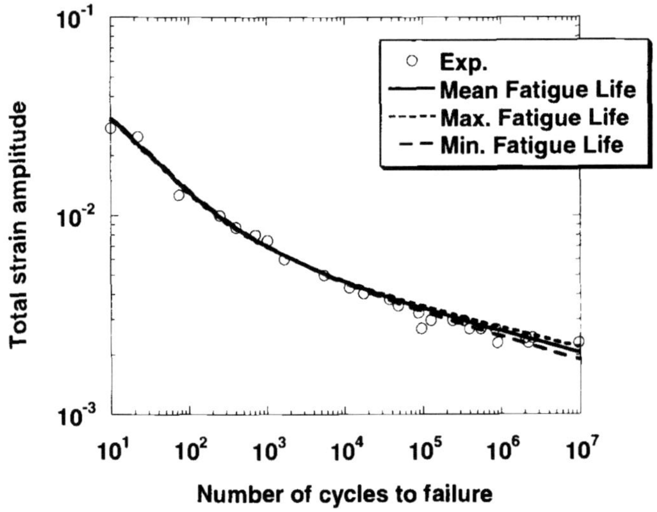

Figure 3 – *Confidence range for the strain-life curve coefficients.*

Regarding to the short crack model, parameters that were allowed to vary are the loading stress, the notch geometry (i.e., K_T that was estimated with different geometries, after taking into account machining tolerances), the initial and final crack size, the crack deflection, the stress-strain curve coefficients and the crack propagation law coefficients.

For the local strain approach, parameters that are allowed to vary are the strain-life and stress-strain curves coefficients, the notch stress concentration factor and the stress input to the specimen.

In order to facilitate the identification of coefficients such as strain-life or stress-strain curves, a graphical user interface (GUI) was developed. The software avoids fastidious data manipulations using spreadsheet software. Experimental points can be selected or unselected (by graphical picking), and the calculation of coefficients is immediately available. Figure 4 shows a screenshot of this function for the stress-strain curve, and Figure 5 for the strain-life one.

It is assumed that variations in prediction due to each input parameter variation are cumulative, and the scatter band is obtained in adding each contribution.

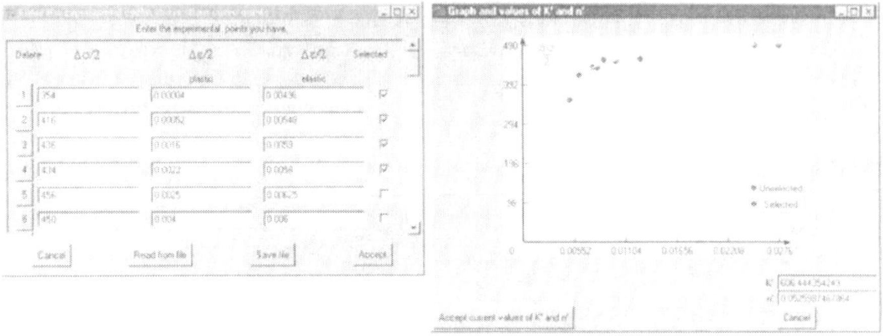

Figure 4 – *Graphical determination of K' and n'.*

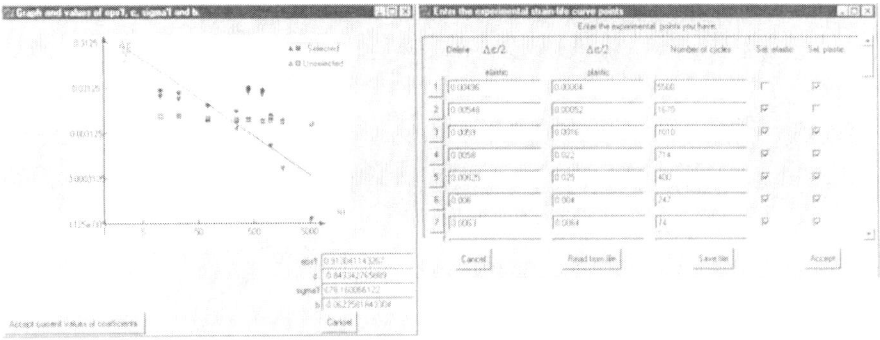

Figure 5 – *Graphical determination of ε'f, c, σ'f and b.*

Predictions: Case Studies

Al 2024-T351, Constant Amplitude Loading

Using experimental results concerning notched components of Al 2024-T351 (K_T= 2.75) under constant amplitude loading (R=0.1), the short crack model was used to evaluate the crack initiation life. For this loading case, no load interaction is expected, and the assumption was made that short cracks propagate without any crack closure, and a power law relating $\frac{da}{dN}$ and ΔK_{eff} was used (intrinsic crack growth law, see Table 1 for the coefficients). Results of predictions, compared with experimentally measured crack initiation lives are shown in Figure 6. Furthermore, each input parameter such as intrinsic material characteristics, mechanical parameters (crack geometry, notch stress concentration parameter...) and loading conditions (remote stress amplitude) were allowed to vary within their confidence range. This led to the prediction of a scatter band, describing quite accurately the experimental scatter.

Table 1 – *Intrinsic Crack Propagation Law:* $\frac{da}{dN} = C.\Delta K_{eff}^{m}$ *Coefficients for the Al 2024-T351.*

ΔK[1] value, MPa.√m	C	m
$\Delta K \leq 2$	1.3333×10^{-11}	7.229
$2 \leq \Delta K \leq 4$	3.3×10^{-10}	2
$\Delta K \geq 4$	1.5×10^{-11}	4.1

[1] ΔK for short cracks corresponds to ΔK_{eff} for long cracks.

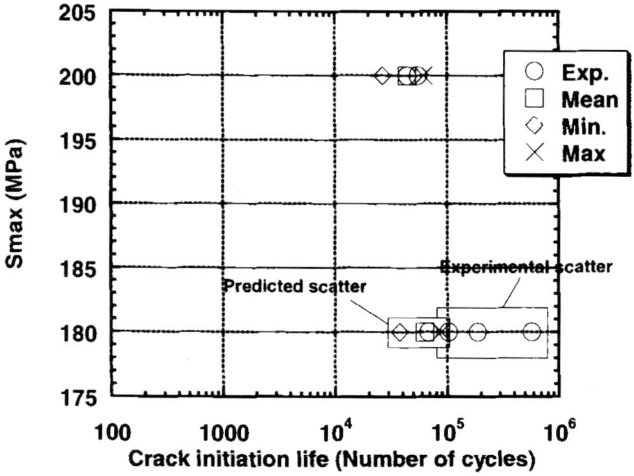

Figure 6 - *Comparison between predicted and experimental crack initiation life, R=0.1, Al 2024-T351, K_T=2.75 ('short crack model').*

The results obtained with a conventional strain life method are given in Figure 7, whereas results given by the 'short crack model' are shown in Figure 6. Both these approaches give conservative predictions as can be seen from Figures 6 and 7, nevertheless, the predictions with the short crack model give results much close to the experimental data. The local strain predictions seem to be overconservative.

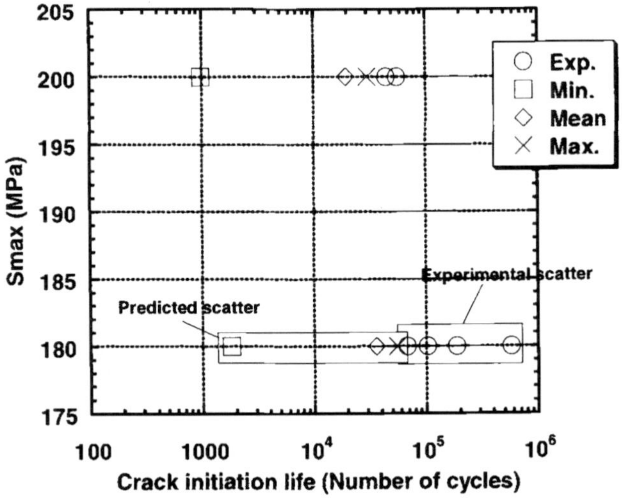

Figure 7 – *Comparison between predicted and experimental crack initiation life, R=0.1, Al 2024-T351,* K_T*=2.75 (local strain approach, Morrow correction).*

Al 2024-T3, Variable Amplitude Loading

Test results using the same notched specimens submitted to a civil aircraft loading spectrum, representing 1,000 flights, each composed of a taking off, a cruising regime, and landing phases (leading to 532,036 cycles per block) are next analyzed. The predictions obtained with the short crack model are shown in Figure 8 whereas those made with the local strain approach and Smith-Watson-Topper correction are shown on Figure 9. Estimations made with the Morrow corrections are not shown for sake of brevity, but were less accurate.

The predictions made with the short crack model are obviously more accurate than those made with the local strain approach because they are generally closer to the experimental initiation lives. At lower stress levels, the short crack model is more accurate than the local strain approach, and the predicted scatter has an acceptable width. The local strain approach gives unconservative results, and the predicted scatter band is about four times larger than experimentally measured.

At higher stress levels, no information was available about the experimental scatter, but the lower bound predicted by the local strain approach is closer to the experimental measure than the lower bound of the short crack model prediction. The fact that the short crack model can be unconservative at higher stress levels will be discussed later.

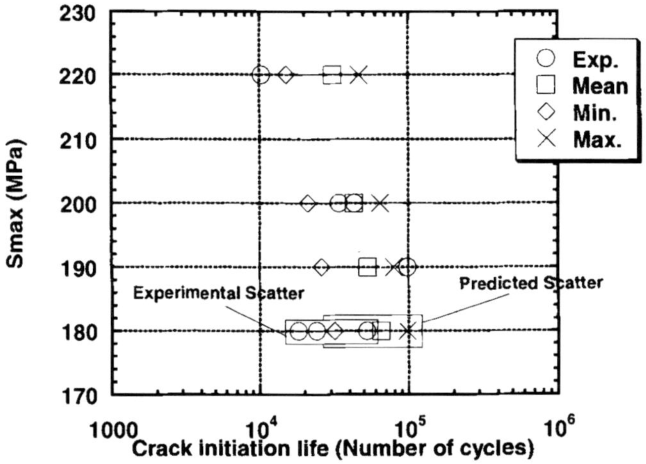

Figure 8 – *Comparison between predicted and experimental crack initiation life, spectrum loading, Al 2024-T351, K_T=2.75 ('short crack model').*

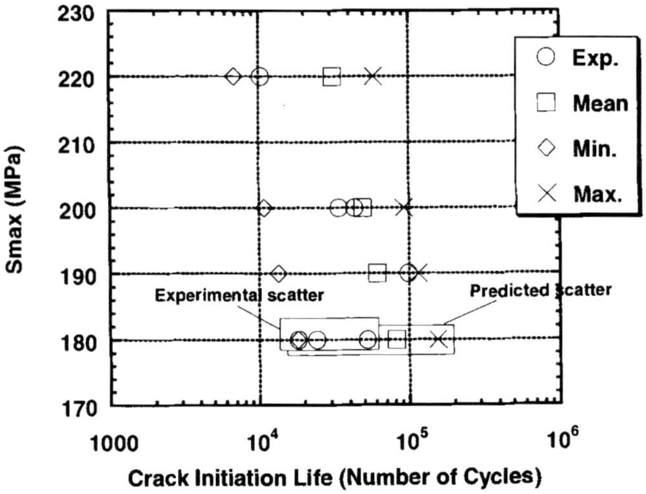

Figure 9 – *Comparison between predicted and experimental crack initiation life, spectrum loading, Al 2024-T351, K_T=2.75 (local strain approach, Smith-Watson-Topper correction).*

Al 7075-T6, Constant Amplitude Loading, Negative Stress Ratio

Notched specimens (K_T=2.3) were submitted to constant amplitude loading with negative stress ratios (-1, -3 ,-5.3) [*12*]. For such loading cases, the previously used assumption (no closure for short cracks) led to overconservative predictions, since it involved the compressive part of the cycle in the crack propagation. Thus a constant 'local opening stress' was determined by fitting predicted and experimental initiation life. The opening ratio U defined by Equation (6) was found to be linear with the local maximum stress, whatever the stress ratio.

The crack initiation life (up to a crack size of 300 μm) in notched specimen made of Al 7075-T6 (K_T=3.2) as studied by Newman et al. [*13*]) was determined with the 'short crack' model. The predictions made by Newman et al. [*13*] are compared in Figure 10 with those obtained with the 'short crack model', using closure expression provided by Pang et al. [*10*] and the closure evolution defined above. The predictions obtained with the 'short crack model' were added to the initial figure [*13*].

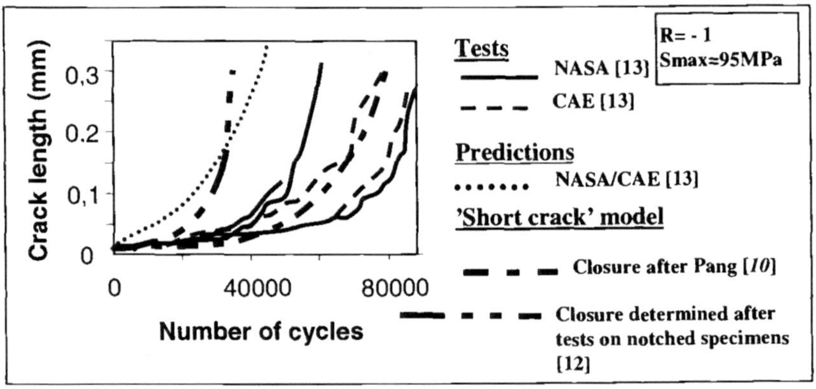

Figure 10 – *Comparison between crack initiation life prediction (a_i=0.3 mm) by the FASTRAN program [13], and the 'short crack model'. Modified after [13].*

Newman et al. [*13*] explained that the predictions obtained with the FASTRAN program are accurate, although they gave lifetimes that are half the experimental ones. The difference has been associated with a possible environmental effect due to the replicating technique used to monitor crack growth. Predictions made with the 'short crack' model (and the crack closure evolution determined with tests made by Nicolas [*12*]) are quite close to those made with the FASTRAN program.

Summary of Case Studies

In Table 2, different cases studies using the short crack approach are described, bringing out the validity of this method for most of the cases studied.

Table 2 – *Cases studies summary.*

Material	Loading conditions	Geometry	Closure hypothesis	Predictions quality (Short crack model)	Observations
Al 2024-T3	CAL, R=0.1	SENT specimen K_T=2.75	No closure	accurate (see Figure 6)	
	VAL (45,094 cycles per block)	SENT specimen K_T=2.75	No closure	accurate	Slightly non conservative at higher stress level
	VAL (532,036 cycles per block)	SENT specimen K_T=2.75	No closure	accurate (see Figure 8)	Slightly non conservative at higher stress level
	Periodic overloads	SENT specimen K_T=2.75	No closure	inaccurate	Over conservative : does not take into account for the beneficial effect of overloads
Al 7075-T6	CAL, R=-5.3,-3,-1	Notched specimen, K_T=2.3	No closure	inaccurate	Does not take account for the amount of closure due to the compressive part of the cycle
Al 7075-T6	CAL, R=-1, S_{max}=95 MPa	SENT specimen, K_T=3.2 (Newman [*13*])	Evolution with σ_{max} determined with previous tests results[1]	accurate	Using a local opening law as a function of the maximum local stress
Al 7075-T6	CAL, R=-1, S_{max}=95 MPa	SENT specimen, K_T=3.2 (Newman [*13*])	Chong-Myong Pang [*10*]	Unconservative (see Figure 10)	Using a $U \propto \Delta K_{eff}$ law [*10*].

[1] The crack closure evolution was determined using previous tests results, by fitting the predictions to the experimental results, using a constant crack closure level for the whole crack life.

Discussion

Advantages and Drawbacks of the Two Methods Used

Strain-Life Approach – For this approach, the present study confirms that the prediction accuracy is strongly dependent upon the accuracy of the parameters representing the strain-life relationship of the material, such as ε'_f, c, σ'_f and b. These parameters cannot be unambiguously defined, especially in the plastic part of the curve (ε_f' and c) as shown in Figure 11, where the studied material exhibits a two-sloped relationship between $\Delta\varepsilon_p$ and N_f.

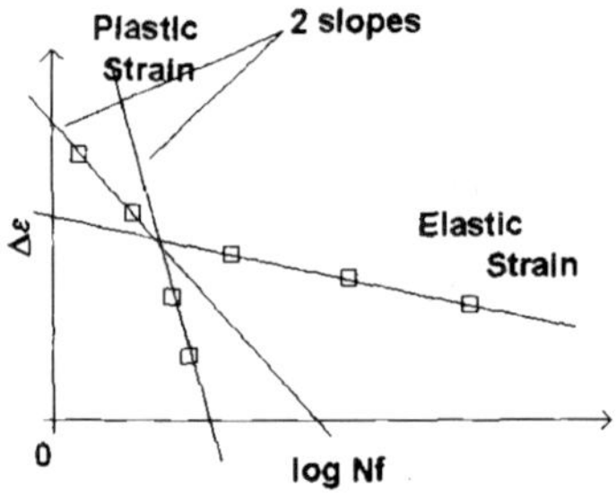

Figure 11 - *Typical behaviour in the low-cycle fatigue part of the strain-life relationship: double slope.*

A disturbing point concerns the identification of parameters representing the cyclic properties of the material such as *K'* and *n'*. Brennan [*14*] has shown, for example, that most of the coefficients used to fit the strain-life curve used in the literature are not consistent with the coefficients of the stress-strain curve of the material. Moreover, the number of smooth specimens tests necessary to determine accurately all these parameters can be numerous and time consuming. Especially, low-cycle fatigue tests in the plastic part of the curve require very careful design and alignment of the fatigue system, as advised by the ASTM Recommended Practice for Constant-Amplitude Low-Cycle Fatigue Testing (E 606).

Another drawback of the local strain approach is the fact that there is no unambiguous choice of the mean stress correction to be used, since the accuracy of the predictions change with the correction chosen for the mean stress effect (among Morrow, Manson-Halford and Smith-Watson-Topper corrections). An example is shown in Table 3, where notched components of K_T=2.75 are submitted to a variable amplitude loading at the same maximum stress of 180 MPa. It can be seen here that the predicted lives can vary by a factor of 2, depending upon the mean stress correction used. It should be mentioned that the Manson-Coffin model does not involve any mean stress correction.

Table 3 – *Comparison between estimations made with Manson-Coffin, Manson-Halford, Morrow, and Smith-Watson-Topper corrections (mean input parameters).*

Correction for the mean stress effect[1]				Experimental life
Smith-Watson-Topper	Morrow	Manson-Halford	Manson-Coffin	
82,000 flights	197,000 flights	158,895 flights	648,000 flights	18,133 to 52,336 flights

[1] Predictions made for K_T=2.75 under loading spectrum, S_{max}=180 MPa (Al 2024-T3).

Short Crack Model - For the short crack model, no strain-life curve data is necessary, because there is no damage summation as in the local strain approach. The only experimental data needed are the cyclic stress-strain curve and the intrinsic crack propagation law. The former can be obtained with a single specimen test (hourglass shape specimen), which can be submitted to several stress levels in order to draw the cyclic stress-strain response of the material (that will give the Ramberg-Osgood equation coefficients K' and n'). The crack growth data are easily obtained by crack propagation measurements made according to the ASTM Test Method for Measurement of Fatigue Crack Growth Rates (E 647). But closure corrections are required to determine the closure free long crack growth relationship. These measurements are delicate to make and as such, there is no unambiguous method for determining this intrinsic crack growth curve. An acceptable approximation is the use of constant K_{max} and high R ratio test to determine the intrinsic curve $\frac{da}{dN} \Leftrightarrow \Delta K$.

Unfortunately, these advantages have to be moderated, because a few drawbacks remain. Several questions concerning short cracks at notches are indeed still open to discussion. Firstly, the evaluation of the adequate crack driving force parameter, since linear elastic fracture mechanics (LEFM) breakdown is widely recognized for such short cracks. Unless using time consuming calculations, such as finite element method (FEM) coupled with severe assumptions regarding to the cyclic material behavior, for a slightly complicated geometry, no direct method exists.

The short crack propagation behavior also is not well defined at this time. As far as the so-called 'short crack effect' is well known and recognized now, no consensus exists regarding this particular behavior of short cracks, when compared with longer cracks. This point joins the LEFM breakdown, as the discrepancies between short and long cracks are sometimes explained this way: the fact that parameters such as ΔK are not applicable to short cracks induces the observed discrepancy on propagation curves $\frac{da}{dN} \propto \Delta K$ between short and long cracks (remember the ASTM specification regarding to the LEFM applicability $\frac{a}{r_p} > 50$ with a being the crack length and r_p the crack-tip plastic zone [*15*]). The most frequent explanation for the particular behavior of short cracks remains a transient state of closure. Short cracks of near zero size are believed to propagate without any crack closure (that reduces the crack driving force after Elber's theory), and the crack closure level is assumed to increase with the crack length, reaching the long crack stationary closure level when short and long cracks propagation curves join together [*16*]. This opinion is widely expressed in the literature and has been observed for numerous metallic materials. In this case, only a correction for closure of the crack driving force parameter (such as ΔK) is necessary and accurate correlation of short and long cracks propagation curves were done using $\Delta K_{eff} = K_{max} - K_{op}$ after Elber's theory [*17*]. Recently, works of Chong-Myong Pang [*10*] regarding the crack closure evolution of short cracks reflect this commonly developed idea, with the advantage of possible extension to variable amplitude loading cases.

Models that take into account this short crack effect are numerous, based on different approaches. One can quote models that are based on interaction between microstructural features and cracks, such as those proposed by Navarro and de los Rios or Tanaka [*18,19*]. Such an interaction is believed to be responsible for the retardation effect observed in the early stages of the short crack life, and these models try to quantify it. Quite accurate predictions of short crack growth have been obtained. The approach developed by Newman is intended to determine the plasticity-induced closure based on a modified Dudgale model. The main advantage of this model is the fact that only one crack propagation curve is needed for any stress ratio: using a constraint state factor α, all stress levels curves are correlated in a unique curve. Even though it is difficult to predict this constraint factor [*20*], predictions obtained with this model are generally accurate [*13*].

The main difficulty of the short crack model remains the stress intensity factor expression, for which analytical expressions exist for some simple geometric configurations but can be a problem for ones that are more complex. This requires FEM calculations (or boundary element method –BEM- calculations) to determine the K calibration. This also involves assumptions regarding the crack aspect ratio evolution, for three-dimensional cases.

As a first approach, it was believed that Kujawski expression [*9*] could give the K values for most of the common crack geometries. Unfortunately, its limits appeared as described below. The main advantage nevertheless is the simplicity of the analytical expression used. Since this expression was a simplification of Newman's and Raju's [*7*] one, its validity was close to results in [*7*].

Furthermore, the corner crack configuration was not studied, and this was remedied as follows: a Q_k factor value of *0.87* (Eq. 4) was determined by curve fitting technique in the Kujawski's expression, as gave us FEM calculations with a notch and specimen geometry within the validity range of Newman's expression. The fitted curve obtained with a Q_k factor of *0.87* is shown in Figure 12, compared with FEM results and Newman's expression. The graphical user interface of the software was further modified to enable input of a crack length *a vs. K* curve, obtained for example with BEM or FEM calculations.

Works of Filippini [*21*] gave us a supplementary reason to use such a procedure, as it shows that the specimen width can strongly modify the stress gradient at the notch. A stress intensity factor expression evaluated for a notch has thus to be modified with the specimen width. Furthermore, only simple cases where analytical exact solutions exist can be directly treated [*21*], otherwise errors can arise out of inaccurate determination of the stress gradient. As the stress intensity factor expression does involve the stress gradient, complex geometric configurations have to be considered with care.

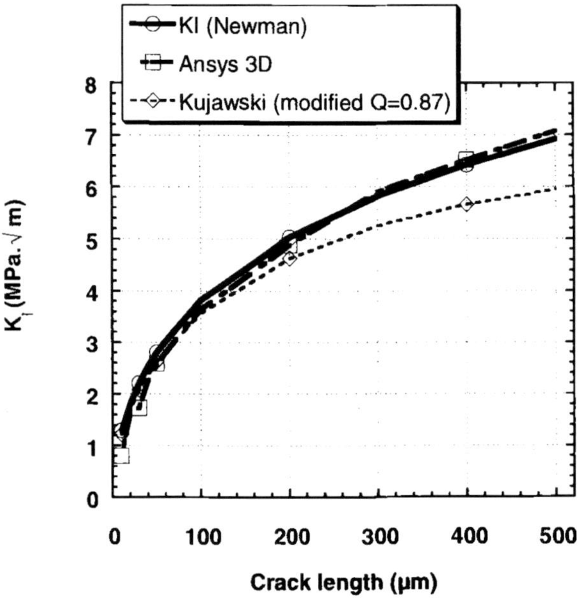

Figure 12 – *Stress intensity factor for a corner crack.*

In the present study, while comparing short crack growth prediction with the experimental lives, it was shown that whereas at low stresses the predictions are quite accurate, at high stresses the predictions are non-conservative. This indicates that the Kujawski expression may not be exact when local plasticity effects are predominant.

This aspect was treated by comparing the stress gradient in the vicinity of the notch predicted by the analytical expression and FEM simulations for two far field stress levels (Figures 13 and 14). From these figures, it can be seen that the Kujawski expression is very accurate at low stress levels while it seems to underestimate the stress gradient when local plasticity effects are predominant.

This might be a possible explanation of why the prediction are non-conservative at high stress levels (see Figure 8), as probably the local ΔK values were underestimated. From this analysis, it appears that it should be necessary to introduce a cut-off local stress at the notch-tip above which the Kujawski expression is no more valid.

Finally, when dimensioning a component submitted to fatigue against failure, no information is available about the crack geometry or crack path deflection. This means all the possible geometrical configurations have to be studied, like embedded penny-shape crack, through-thickness crack, and corner crack.

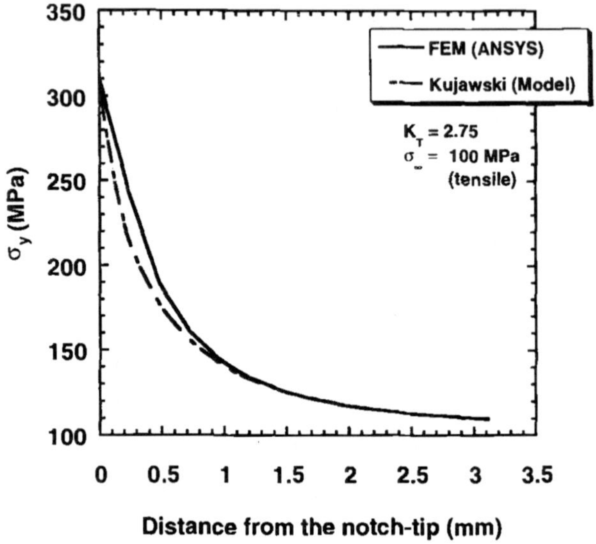

Figure 13 – *Stress gradient ahead of a notch: comparison between FEM simulations (elastic-plastic) and Kujawski's expression as used in the 'short crack model' (small-scale yielding case).*

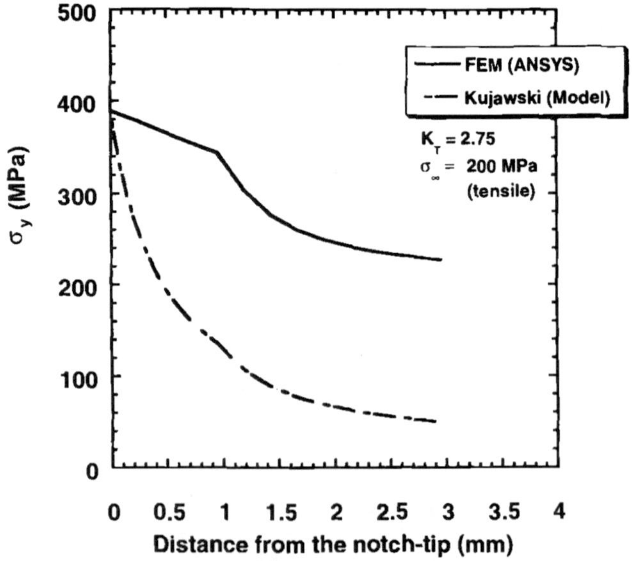

Figure 14 – *Stress gradient ahead of a notch: comparison between FEM simulations (elastic-plastic) and Kujawski's expression as used in the 'short crack model' (large plasticity case).*

The crack orientation can be varied within a reasonable extent depending upon the material. Such geometrical parameters contribute to scatter, that can be nevertheless found in experimental measurements.

Lack of the Model/Software and Improvements

Regarding the stress intensity factor based (short crack) model, the most important point is the determination of the stress intensity factor evolution as the crack grows. Obviously, the incorporation of a simple FEM solver could provide an efficient way to get accurate predictions. Such an approach is used, for example, in software such as FRANC2D and FRANC3D [22]. The first step was done, since such a generic curve of K_I *vs.* crack length (obtained elsewhere) can be incorporated.

The GUI functionality will be enhanced, keeping the same aim: to permit evaluating the confidence range of parameters that are evaluated after experimental data. As described earlier, it is indeed believed that allowing variation of those parameters within their confidence range can lead to the determination of the scatter in actual lives.

In a first approximation, it was assumed that short cracks propagate without any crack closure. A crack propagation law was then used from long crack tests at high stress ratio, as no crack closure is generally assumed for such configurations. Since no consensus exists yet about abnormal short crack propagation behavior, it is believed that including several different corrections for crack closure can provide an efficient answer to engineers, as dimensioning can then be made using the more damaging model.

In typical loading spectra, load interaction effects due to infrequent peak loads may be present. It then becomes important to be able to quantify effects such as crack growth retardation after an overload, or beneficial effect of smaller cycles embedded between larger cycles. In the present, loading spectra that were studied did not exhibit any typical interaction effect, and the model gave quite accurate predictions. Nevertheless, for typical loading such as periodic overloads, the model failed to evaluate crack initiation life, as it did not take into account the crack growth retardation. This is one of the points to be improved in the future.

In the case of the strain-life approach, the main improvement required concerns the choice of an adequate mean stress correction. Another improvement concerns the development of a non linear damage law, such as the one proposed by Chaboche [*23*].

These future improvements are indicated in Figure 15, which represent an 'ideal' flowchart.

Conclusion

The short crack model has shown its efficiency for several loading cases, especially positive stress ratios and loading spectrum without any loading interaction. For particular situations, where simple assumptions such as 'no closure for short cracks' are obviously

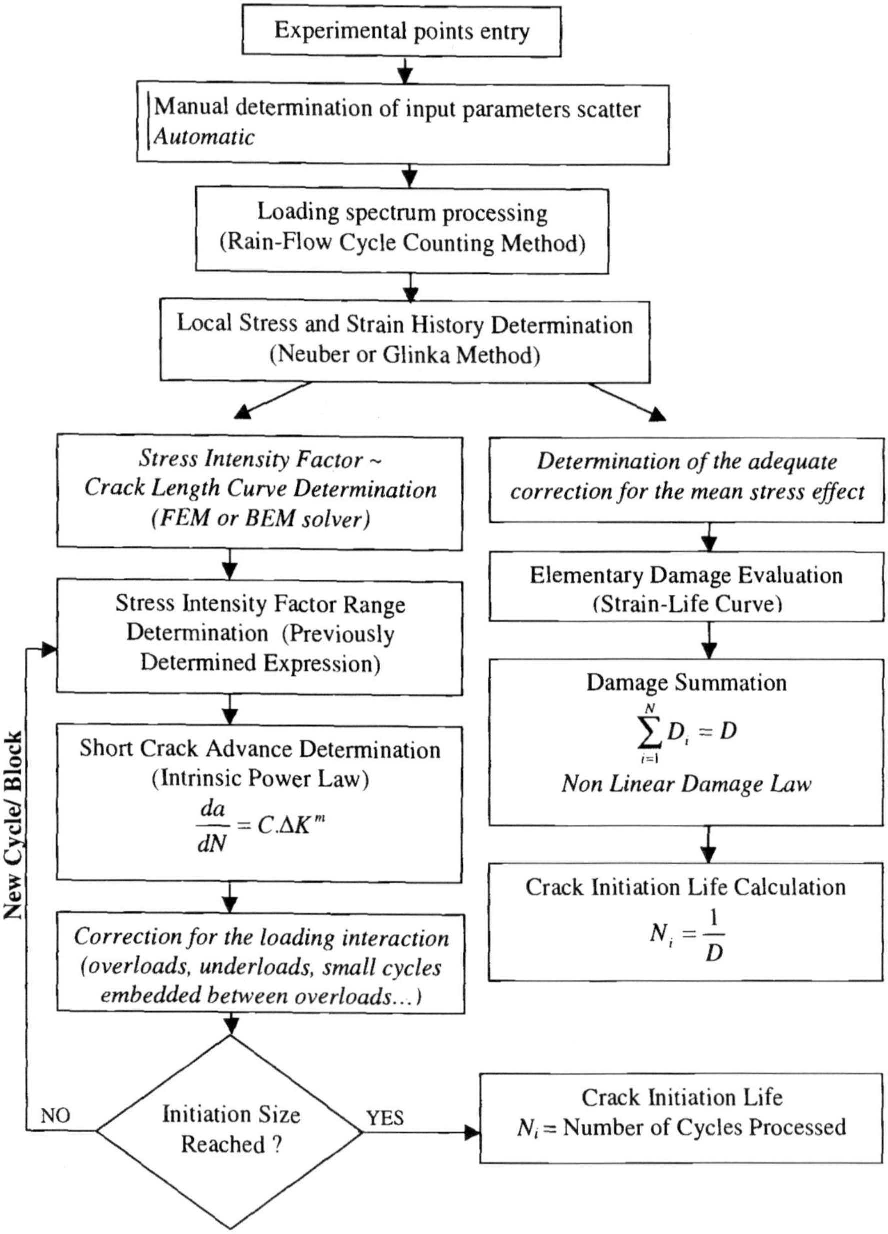

Figure 15 – *Currently available model/software features and future improvements (in plain text: available features; in italic text: future -or under development- improvements).*

inadequate (negative stress ratios), it was found that an equivalent crack closure level can lead to adequate predictions.

It was shown that allowing variations of input parameters within their confidence range can lead to the prediction of the experimental scatter. The main advantage of such an approach is the low cost in term of calculation time, since it leads to an acceptable prediction of the experimental scatter with only few calculations.

References

[1] Dowling, N. E., "A Review of Fatigue Prediction Methods" *Passenger Car Meeting and Exposition*, Dearborn, Michigan, October 1987.

[2] Ye Du-Yi and Wang De-Jun, "A New Approach to the Prediction of Fatigue Notch Reduction Factor K_f" *International Journal of Fatigue*, Vol. 18, N° 2, 1996, pp. 105-109.

[3] Glinka, G., "*Fatigue Life Predictions of Notched Components: a Crack Initiation*" University of Waterloo, Ontario, Canada, November 1990.

[4] Glinka, G. and Kam, J.C.P., "Rainflow Counting Algorithm for Very Long Stress Histories" *International Journal of Fatigue*, N° 3, 1987, pp. 223-228.

[5] Glinka, G., W. Ott, and Nowack, H. "Elasto-Plastic Plane Strain Analysis of Stresses and Strains at the Notch Root" *Journal of Engineering Materials and Technology*, Vol. 110, 1988, N° 3, pp. 195-204.

[6] Ranganathan, N., Morel, F., and Nadot, Y., In *Engineering Against Fatigue*, Lindley, Smith, Tomkins, Beynon, and Brown, Eds., 1999, pp. 183-189.

[7] Newman, J.C. and Raju, I.S., "Stress-Intensity Factor Equations for Cracks in Three-Dimensional Finite Bodies Subjected to Tension and Bending Loads" *Computational Methods in the Mechanics of Fracture*, Atluri, Ed., Vol. 2 in *Computational Methods in Mechanics*, Elsevier Science Publ. Co., Inc., 1986, pp. 311-334.

[8] Lukáš, P., "Stress Intensity Factor For Small Notch-Emanated Cracks" *Engineering Fracture Mechanics*, Vol. 26, N° 3, 1987, pp. 471-473.

[9] Kujawski, D., "Estimations of Stress Intensity Factors for Small Cracks at Notches" *Fatigue and Fracture Engineering of Materials and Structures*, Vol. 14, N° 10, 1991, pp. 953-965.

[10] Chong-Myong Pang and Ji-Ho Song, "Crack growth and Closure Behaviour of Short Fatigue Cracks" *Engineering Fracture Mechanics*, Vol.47, N° 3, 1994, pp. 327-343.

[11] Laz, P.J. and Hillberry, B.M., "Fatigue Life Prediction from Inclusion Initiated Cracks" *International Journal of Fatigue*, 1998, pp. 263-270.

[12] Nicolas, Ch., "Etude du Comportement en Fatigue de l'Alliage d'Aluminium 7075-T651 sous Sollicitation en Compression Predominante " *PhD Thesis*, University of Poitiers (France), 1993 .

[13] Newman, J.C. JR, Wu, X. R., Venneri, S.L. and Li, C.G., "Small-Crack Effects in High Strength Aluminum Alloys, A NASA/CAE Cooperative Program", *NASA Reference Publication 1309*, May 1994.

[14] Brennan, F.P., "The Use of Approximate Strain-Life Fatigue Crack Initiation Predictions" *International Journal of Fatigue*, Vol. 16, July 1994, pp. 351-356.

[15] Hussain, K., "Short Fatigue Crack Behaviour and Analytical Models: a Review" *Engineering Fracture Mechanics*, Vol. 58, N° 4, 1997, pp. 327-354.

[16] Bathias, C. and Baïlon, J.P., "La Fatigue des Matériaux", Hermès Ed.

[17] Tanaka, K. and Nakai, Y., "Propagation and Non-Propagation of Short Fatigue Cracks at a Sharp Notch" *Fatigue of Engineering Materials and Structures*, Vol. 6, N° 4, 1983, pp. 315-327.

[18] Navarro, A. and de los Rios, E.R., "A Microstructurally-Short Fatigue Crack Growth Equation" *Fatigue and Fracture of Engineering Materials and Strutures*, Vol. 15, 1988, pp. 383-396.

[19] Tanaka, K., Akiniwa, Y., Nakai, Y. and Wei, R.P., "Modelling of Small Fatigue Crack Growth Interacting With Grain Boundary" *Engineering Fracture Mechanics*, Vol. 24, N° 6, 1986, pp. 803-819.

[20] Newman J.C., "A Crack-Closure Model for Predicting Fatigue Crack Growth under Aircraft Spectrum Loading" *Methods and Models for Predicting Fatigue Crack Growth under Random Loading, ASTM STP 748*, Chang and Hudson Eds., American Society for Testing and Materials, 1981, pp. 53-84.

[21] Filippini, M., "Stress Gradient Calculations at Notches" *International Journal of Fatigue*, Vol. 22, 2000, pp. 397-409.

[22] Computational Fatigue Group, University of Cornell, URL: http://www.cfg.cornell.edu/software/CFG_software.html

[23] Chaboche, J.-L., "A Different Law for Non Linear Cumulative Fatigue Damage" *Materials and Buildings Researches*, Annales de l'ITBTP HS 39, 1977, pp. 117-124.

Measurement and Analysis

M. Khalil,[1] D. DuQuesnay,[2] and T. H. Topper[3]

Prediction of Crack-Opening Stress Levels for Service Loading Spectra

Reference: Khalil, M., DuQuesnay, D., and Topper, T. H., **"Prediction of Crack-Opening Stress Levels for Service Loading Spectra"** *Applications of Automation Technology in Fatigue and Fracture Testing and Analysis, ASTM STP 1411,* A. A. Braun, P. C. McKeighan, A. M. Nicolson, and R. D. Lohr, Eds., American Society for Testing and Materials, West Conshohocken, PA, 2002.

Abstract: The fatigue lives of automotive components subjected to variable amplitude service loading are assumed to be dominated by a process of small crack growth. It is generally accepted that crack closure is responsible for most of the variation in fatigue crack growth rates and fatigue lives. This investigation examined the effect of different types and magnitudes of service loadings on crack closure behavior. Three different Society of Automotive Engineers (SAE) standard service load histories with different mean stresses were applied to notched specimens of a 2024-T351 aluminum alloy. The three spectra are the SAE Grapple Skidder history, which has a positive mean stress, the SAE Log Skidder history which has a zero mean stress, and the inverse of the SAE Grapple Skidder history which has a negative mean stress. A curve of maximum stress in the history versus fatigue life was constructed for each spectrum. The crack-opening stress (COS) levels were measured at frequent intervals in order to capture the behavior of the opening stress for each spectrum, for each of a set of scaled histories with different maximum stress ranges.

A crack growth analysis based on a fracture mechanics approach was used to model the fatigue behavior of the aluminum alloy specimens for the given load spectra and stress ranges. The crack growth analysis was based on an effective strain-based intensity factor, a crack growth rate curve obtained during closure-free loading cycles, and a local notch strain calculation based on Neuber's rule. The COS levels were modeled assuming that the COS follows an exponential build-up formula that is a function of the difference between the current crack opening stress and the steady state crack opening stress of the given cycle unless this cycle is below the intrinsic stress range, or the maximum stress is below zero. However, the build-up only occurs when the crack-opening steady state stress level for the given cycle is higher than the previously calculated crack opening stress. The modeled crack opening stress level was in good agreement with the measured crack opening stress.

[1]Research Assistant, Civil Engineering Department, University of Waterloo, Waterloo, Ontario, Canada
[2]Professor, Mechanical Engineering Department, Royal Military Collage of Canada, Kingston, Ontario, Canada
[3]Professor Emirates, Civil Engineering Department, University of Waterloo, Waterloo, Ontario, Canada

The average of the measured crack opening stresses and those calculated using the model were nearly the same for all the histories examined. When these average crack-opening stresses were used in the life prediction model they gave predictions as good as those obtained by modeling COS on a cycle by cycle basis. In the interest of simplifying the use of COS in design the average COS was correlated with the frequency of occurrence of the cycle reducing the COS to the average level. The use of a COS level corresponding to the one in 200 cycle gave a conservative estimate of average COS for all the histories.

Keywords: Service spectrum, crack opening stress, effective strain intensity factor, steady state crack-opening stress

Introduction

Elber [*1*] attributed the crack closure phenomenon to plasticity induced crack closure at the crack tip. DuQuesnay [*2*] correlated physical measurements of crack closure levels with an empirical model for the steady state crack opening stress S_{op} under constant amplitude loading. Duquesnay's crack closure model has been shown to be a good fit to constant amplitude crack closure data for a wide variety of models. However, most service load histories are variable amplitude histories. Many researchers have documented the effects of variable amplitude loading on crack closure [*3-9*]. The effect of variable amplitude loading is usually illustrated by examining a crack with an imposed overload cycle, tensile or a compressive, in between two constant amplitude blocks.

The application of a tensile overload can cause either an acceleration or a delay in crack growth. A post overload increase in crack closure level and crack growth retardation occurs when the applied overload is less than approximately one half the yield stress of the material [*7, 10, 11*]. An overload of much more than one half the yield stress of the material will decrease the crack closure level and accelerate crack growth. Pompetzki et al. [*12-13*] investigated periodic tensile overloads of yield stress magnitude and found an accelerated fatigue damage. They postulated an interaction damage model, based on crack closure concepts, with an effective stress as the damage parameter. Their model assumed that the crack opening stress was reduced immediately following a high stress overload and that closure then gradually built up to a steady-state level during subsequent cycling.

Applying an underload causes a flattening of the asperities in the crack wake, which decreases the crack closure level and in turn increases the effective stress intensity factor and accelerates crack propagation. Topper and co-workers [*10, 14*] investigated the effect of intermittent compressive underloads on an aluminum alloy and a steel. The reduction in the COS and its recovery following the compressive underloads is documented in their study. Compressive underloads caused a significant decrease in the threshold stress intensity and increased the crack growth rate. They showed that after a compressive underload the crack opening stress of an annealed SAE 1010 steel was decreased and more than 10000 constant amplitude cycles were required for it to return to the stable constant amplitude level. It should also be noted that they found an immediate decrease in the COS following a near yield stress tensile overload or a compressive underload.

The effect of the magnitude of the compressive stress in a fatigue cycle on fatigue crack growth was reported in a subsequent paper by Yu et al. [*15*]. They reported that with an increase in the magnitude of the compressive peak stress in a compression-tension test, the crack propagation rate increased and the threshold stress intensity factor decreased. They also reported that a crack in a notched aluminum alloy 2024 T351 was only closed for part of the stress cycle during compression-compression tests and cracks grew in compression-compression cycling.

From a study of crack closure in a block history with constant amplitude cycles and a tensile overload or underload. Dabayeh and Topper [*16*] proposed a formula to simulate the buildup of crack opening stress after the underload or overload. They found that the crack opening stress drops immediately after an underload or an overload and then increases at an exponentially decreasing rate under the following constant amplitude loading. They also found that the number of cycles needed for the COS to recover to a steady state level increases as the stress difference between the post underload or overload crack opening stress and the steady state crack opening stress increases.

This research is aimed at providing crack closure inputs for modeling fatigue damage or crack growth in a specimen under service loading spectra. Little data are available on this topic. Most of the investigations of crack closure were conducted on through cracks in smooth or notched specimens under constant amplitude loading or constant amplitude loading followed by an overload (variable amplitude loading). The crack opening stress behavior of cracked specimens under service loading spectra is modeled and the accuracy of the model is evaluated experimentally using measured crack closure data.

Material and Experimental Methods

The material used in this study is 2024-T351 high strength aluminum alloy, which is primarily used in the aircraft industry. The chemical composition and mechanical properties of the material are given in Tables 1 and 2, respectively. All testing was carried out on round threaded specimens with a flat gauge length profile. The geometry and dimensions are shown in Figure 1. An edge notch of 0.6 mm diameter was machined into one side of the specimen, at mid length as shown in (Figure 1). Using a notched specimen allows the stresses to exceed the material yield stress at the notch root without buckling or tensile yielding of the whole specimen. This notch size was small enough that, once initiated, the crack rapidly grew out of the zone of influence of the notch. The specimens were fabricated from bars of the material, with the loading axis of each specimen parallel to the direction of rolling. The gauge section of specimens were roughed out on a lathe then finished by progressively shallower cuts. The smooth and notched specimens were prepared in accordance with ASTM standard "Recommended Practice for Constant Amplitude Low-Cycle Fatigue Testing" (E 606-80). The standard recommends hand polishing of the specimens in the loading direction using progressively finer grades of emery paper that varies between 400 to 600. To enhance crack closure observations a final polish using aluminum metal powder were applied with a very fine cloth.

The crack opening stress was measured using a 900x power short focal length optical video microscope at given cycles before and after an overload occurs. A vernier with an accuracy of 0.01 mm was attached to the microscope to measure the crack length and to assure that the crack tip is beyond the notch stress field. A digital process control computer was used to output the three service loading spectra in the form of a sinusoidal loading wave. All fatigue tests on the 2024-T351 aluminum alloy specimens were conducted under load control at a frequency between 1 Hz and 80 Hz, depending on the stress amplitude, using FLEX control software [*17*].

The procedure for measuring the crack opening stress was to stop the test at the maximum stress of the required cycle and then to decrease the load manually while observing the crack tip region in the monitor attached to the optical video until the crack surfaces start to touch each other. Two sets of readings were recorded for each given cyclic stress level, and the average was calculated.

Table 1 – *Chemical composition (percentage by weight).*

Alloy	C	Fe	Cu	Mn	Mg	Cr	Zn	Ti
2024-T351	0.50	0.50	4.35	0.60	1.50	0.10	0.25	0.15

Table 2 – *Measured mechanical properties of Al 2024-T351.*

Mechanical Properties	Units	Magnitude
Modulus of Elasticity	MPa	72400
Tensile Yield Stress (0.2% offset)	MPa	360
Cyclic Yield Stress (0.2% offset)	MPa	450
Ultimate Tensile Stress	MPa	466
True Fracture Stress	MPa	623
Area Reduction	%	35

Experimental Results

Three different commonly used automotive service spectra with different mean stresses were applied to notched specimens of a 2024-T351 aluminum alloy. The three spectra are the torsion channel of the Society of Automotive Engineers SAE GKN Grapple-Skidder (GSK) history, which has a positive mean stress, the reversed SAE GKN Grapple- Skidder (IGSK) history, which has a negative mean stress, and the cable channel of the SAE Log-Skidder history (LSK), which has a zero mean stress. The SAE GKN history was supplied in the form of normalized sequential peak and valley points with a maximum value of 318 and a minimum value of -238 and contains 41124 reversals. The history is shown in Figure 2. The SAE GKN history is multiplied by -1 to produce a history with a negative mean stress. This history is shown in Figure 3. The LSK history was also supplied in the form of normalized sequential peak and valley points with a maximum value of 7.3 and a minimum value of -7.7 and contains 62326 reversals. The LSK history is shown in Figure 4. The values of the peaks of each history were scaled up or down in magnitude to obtain different test stress levels. The fatigue life for each specimen was recorded together with the maximum absolute value of the stress range throughout the scaled histories.

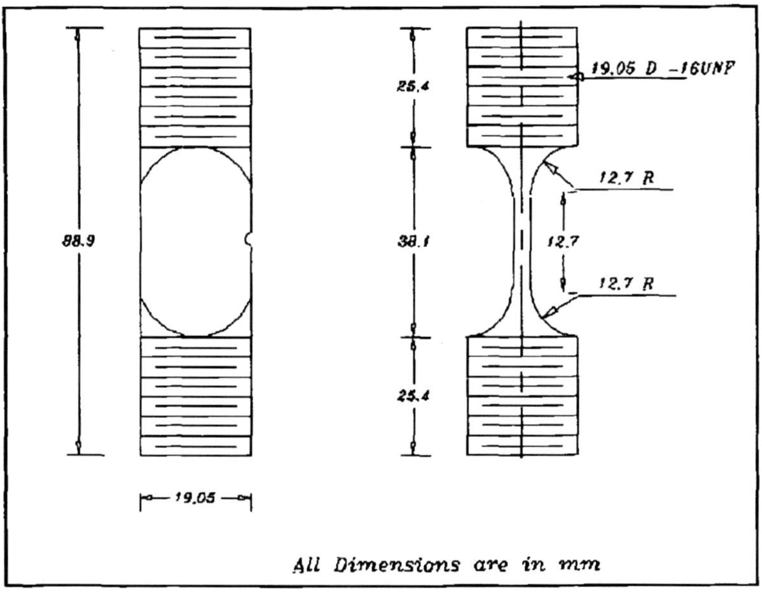

Figure 1 - *Specimen geometry.*

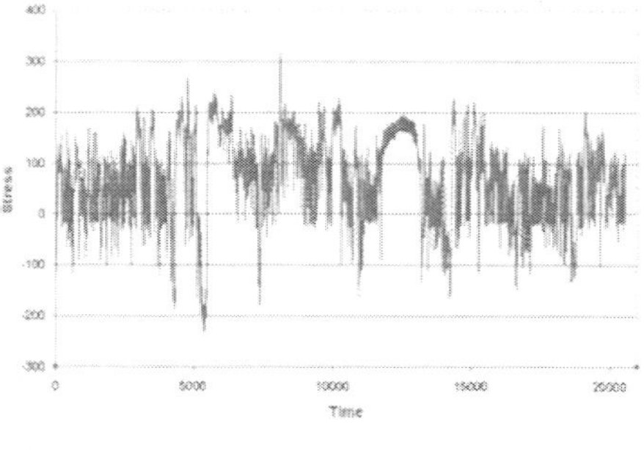

Figure 2 - *Grapple skidder history.*

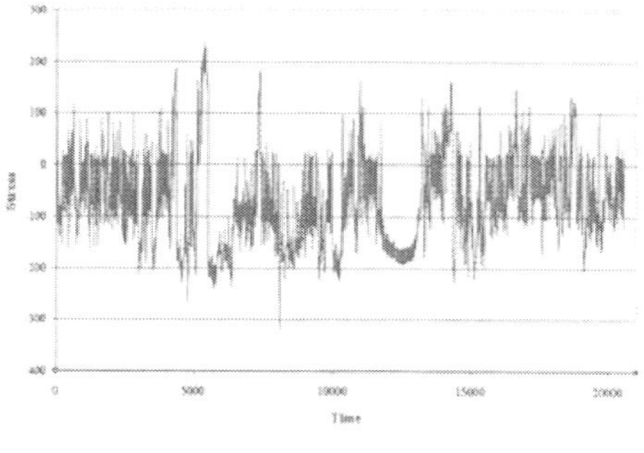

Figure 3 - *Inverse grapple skidder history.*

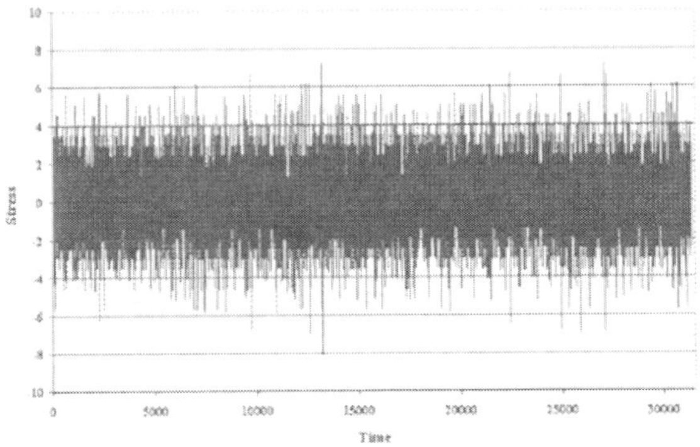

Figure 4 - *Log skidder history.*

Crack Opening Stress Model

Dabayeh and Topper [*16*] measured crack opening stress changes for various levels of tensile and compressive overloads followed by constant amplitude cycles having a variety of R-ratios. The crack opening stress after being reduced by the overload increased to its steady state level in an approximately exponential manner. They showed that when normalized all the closure stress versus cycles, data fell onto a single curve. However, the application of their relationship to complex load histories is complicated. The present authors have applied a simpler relationship suggested by Vorwald and Seeger [*18*] relating the change in crack opening stress in a given cycle to the difference between the current opening stress S_{cu} and the steady state opening stress S_{ss}

$$\Delta S_{op} = m(S_{ss} - S_{cu}) \qquad (1)$$

where ΔS_{op} is the instantaneous increase in crack opening stress and m is a material constant produced by fitting equation (1) to the COS build-up measurements for 2024-T351 aluminum alloy given by Dabayeh and Topper [*16*]. A value of m equal to 0.003 gives a good fit to Dabayeh and Topper's measured COS data.

Using Equation 1 the COS levels were modeled assuming that the COS for a given cycle decreases to the constant amplitude steady state level for that cycle if this steady state crack opening stress is lower than the current opening stress. Otherwise it follows the exponential build-up formula of Equation 1 unless the cycle is below the intrinsic stress range, or the maximum stress is below zero.

Crack Growth Analysis

A crack growth analysis based on a fracture mechanics approach was used to model the fatigue behavior of the aluminum alloy specimens for the given load spectra and stress ranges. The crack growth analysis was based on an effective strain-based intensity factor, a crack growth rate curve obtained during closure-free loading cycles, and a local notch strain calculation based on Neuber's rule [*19*].

Experimental Measurements

The three spectra were scaled to different ranges. The limit of these ranges represent the maximum and minimum stresses that can be applied for a fatigue crack to grow out of a notch without large scale plasticity in the specimen on one hand or an extremely long life on the other hand.

Results for the SAE Grapple-Skidder Load History

The torsion channel of the Grapple-Skidder spectrum which has a positive mean stress was applied to the notched 2024-T351 aluminum alloy. The spectrum was scaled to observe the crack opening stress behavior of the material at three different stress levels. These stress levels are: a maximum nominal stress range of 680 MPa, a minimum nominal stress range of 440 MPa, and an intermediate nominal stress range of 564 Mpa. The crack opening stresses for the 2024-T351 aluminum alloy for the three different stresses are shown in Figures 5 to 7, respectively. The figures show the nominal applied spectrum for each stress range as mentioned above, the calculated crack opening stresses using the crack opening stress model, and the experimentally detected crack opening stresses using the 900 power short focal length optical video microscope. Figure 8 is an expanded view of a portion of Figure 5 that shows the modeled and the experimental crack opening stress. As expected, the crack opening stress decreases when the specimen is subjected to a large overload, then starts to build-up again during subsequent smaller cycles. The large tensile overload causes the crack to stretch open and leads to a drop of the closure level. On the other hand, a compressive overload causes yielding in the wake of the crack and flattening of the crack asperities, which also reduces the crack closure level. The smaller intermittent load cycles between overloads cause the crack tip to grow into the plastically deformed material on the crack tip created during the overload cycle, which results in a buildup of the crack tip closure level.

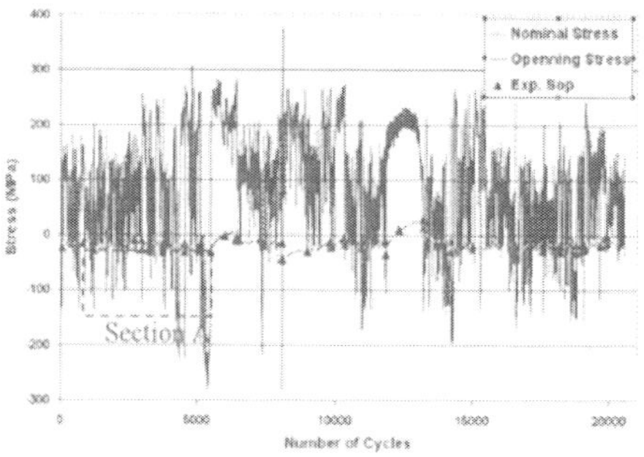

Figure 5 - *GSK maximum stress 680 MPa.*

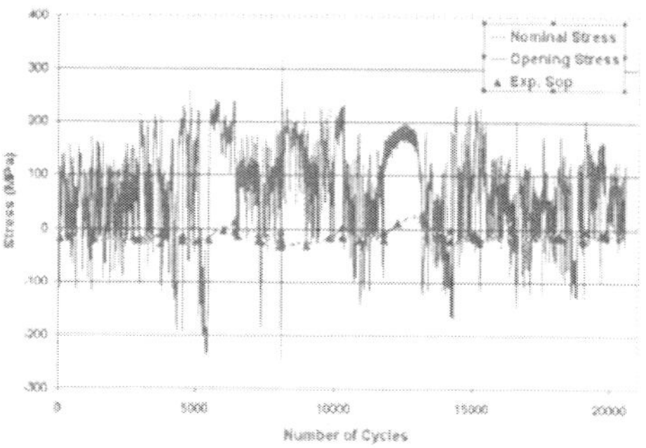

Figure 6 - *GSK maximum stress 564 MPa.*

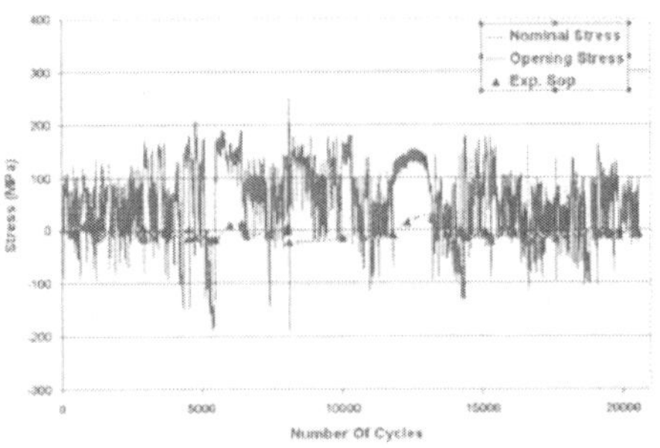

Figure 7 - *GSK Maximum stress 450 MPa.*

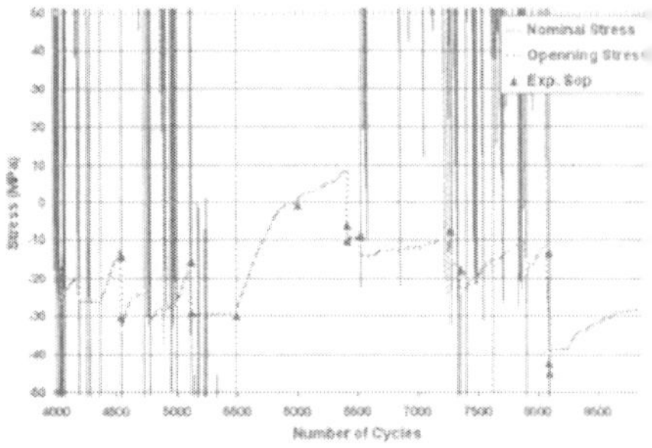

Figure 8 - *Section A.*

The calculated S_{op} and the experimentally measured S_{op} were presented in Table 3 showing the maximum, minimum and the mean opening stresses for the three stress ranges respectively. The calculated S_{op} values were in good agreement with the experimentally determined S_{op} values. Figure 9 shows a plot of the estimated fatigue lives for AL 2024-T351 versus the max stress range under the Grapple-Skidder spectrum and the lives predicted using the COS model in the fatigue crack growth program. The estimated fatigue lives are in good agreement with the experimental observations.

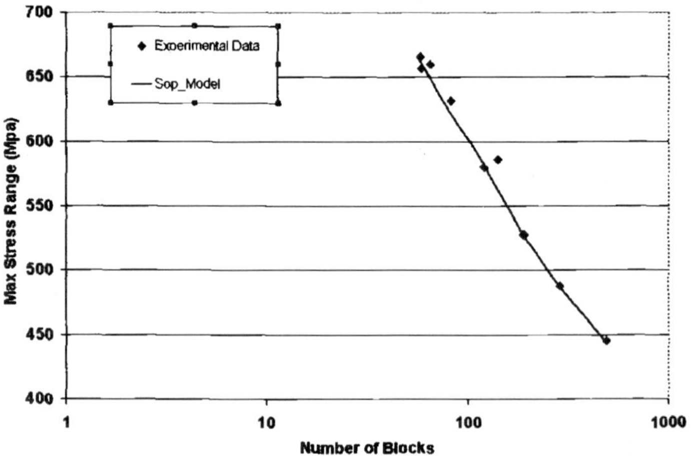

Figure 9 - *Fatigue life versus maximum stress (GSK).*

Table 3 - *Calculated and experimentally measured crack opening stress.*

History	Stress Ranges	Calculated Sop (MPa)			Experimentally Measured Sop (MPa)		
		Maximum	Minimum	Mean	Maximum	Minimum	Mean
Grapple Skidder	680	30	-41	-16.1	29	-45	-16.8
	564	28	-35	-12.8	29	-33	-11.8
	450	24	-28	-6.7	25	-24	-4.7
Log Skidder	677	-9	-39	-27.9	-11	-43	-28
	450	0	-32	-19.3	11.5	-31	-21
Inverse Grapple Skidder	680	3.3	-37	-13	8.4	-34	-16
	450	2.5	-29	-11.9	5.7	-29	-8

Results for the SAE Log Skidder Load History

The cable channel of the Log Skidder spectrum, which had a zero mean stress, was applied to the notched 2024-T351 aluminum alloy. The spectrum was scaled to two different stress levels. These stress levels were: A maximum nominal stress range of 677 MPa and a minimum nominal stress range of 450 MPa.

The crack opening stresses for the maximum and intermediate stress ranges are shown in Figures 10 and 11 respectively. The figures show the nominal applied spectrum for maximum stress ranges of 677 MPa and 450 MPa, the calculated crack opening stresses using the crack opening stress model and the experimentally measured crack opening stresses using the 900 power short focal length optical video microscope. Figure 12 is an expanded view of a portion of Figure 10 that shows the modeled and experimental crack opening stress. Table 3 gives the maximum, minimum and mean S_{op} for the calculated and experimentally determined COS at the two stress levels respectively. The calculated S_{op} values were close to the experimentally determined one. Figure 13 shows the estimated fatigue lives for AL 2024-T351 versus the max stress range under the Log Skidder spectrum. The estimated fatigue lives are in good agreement with the experimental values.

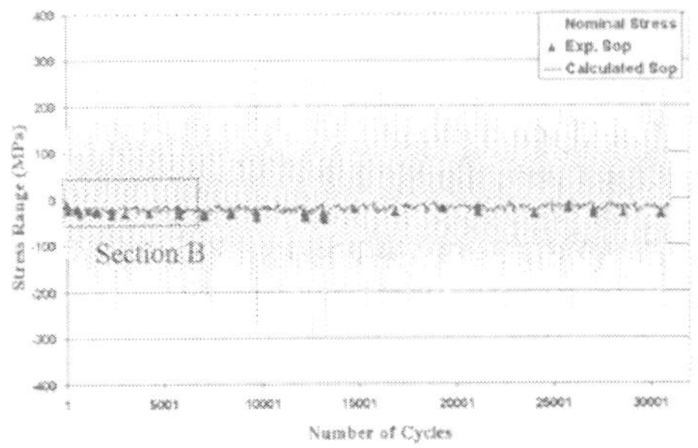

Figure 10 - *LSK maximum stress 677 MPa.*

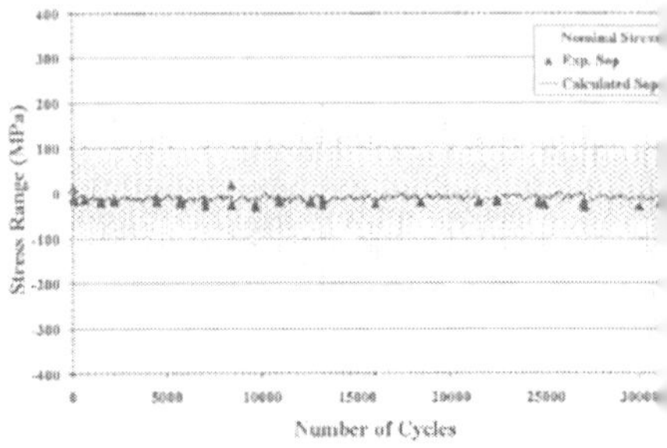

Figure 11 - *LSK maximum stress 450 MP*

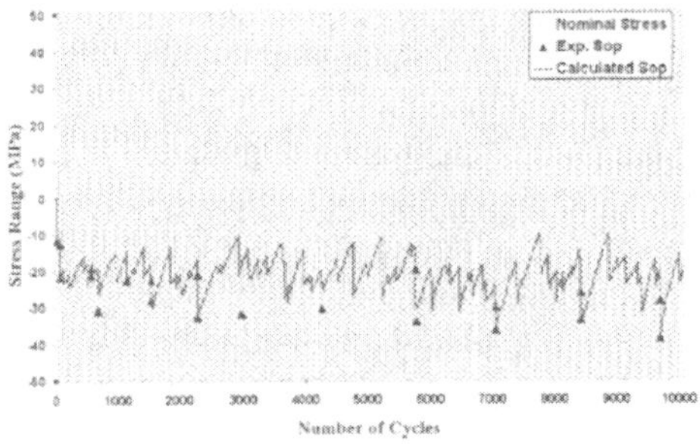

Figure 12 - *Section B.*

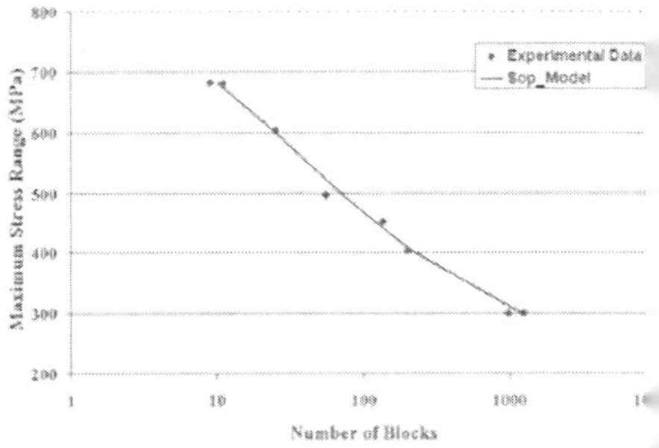

Figure 13 - *Fatigue life vs. maximum stress (LSK*

Results for the Inverse Grapple Skidder Load History

The inverse of the torsion channel of the Grapple Skidder spectrum, which had a negative mean stress, was applied to the notched 2024-T351 aluminum alloy. The spectrum was scaled to two different stress levels. These stress levels are: A maximum nominal stress range of 680 MPa and a minimum nominal stress range of 450 MPa.

The crack opening stresses for the maximum and minimum stress ranges are shown Figures 14 and 15 respectively. The figures show the nominal applied spectrum for maximum stress ranges of 680 MPa and 450 MPa, the calculated crack opening stresses using the crack opening stress model and the experimentally measured crack opening stresses using the 900 power short focal length optical video microscope. Figure 16 is a expanded view of a portion of Figure 14 that shows the modeled and experimental crack opening stress. Table 3 gives the maximum, minimum and mean S_{op} for the calculated and experimentally determined COS at the two stress levels respectively. The calculated S_{op} values were close to the experimentally determined ones. Figure 17 shows the estimated fatigue lives for AL 2024-T351 versus the max stress range under the inverse Grapple Skidder Spectrum. The estimated fatigue lives are in good agreement with the experimental values.

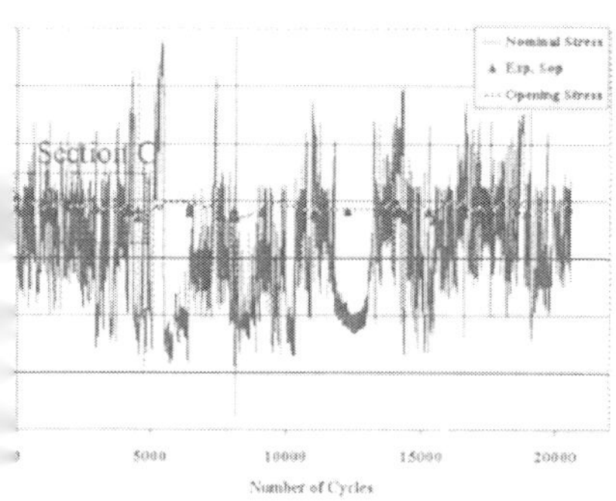

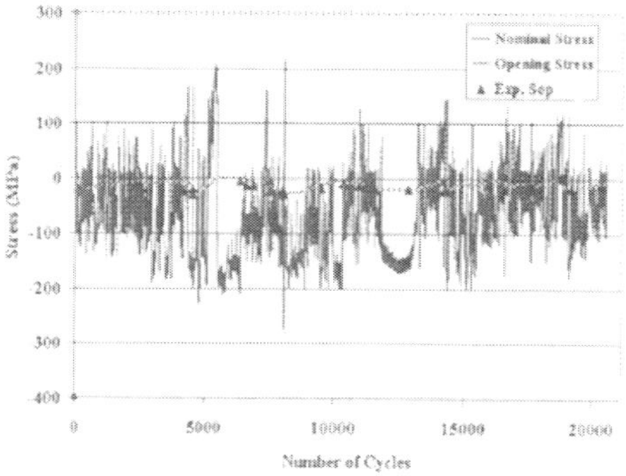

Figure 14 - *IGSK maximum stress 680 MPa.* Figure 15 - *IGSK maximum stress 450 MPa.*

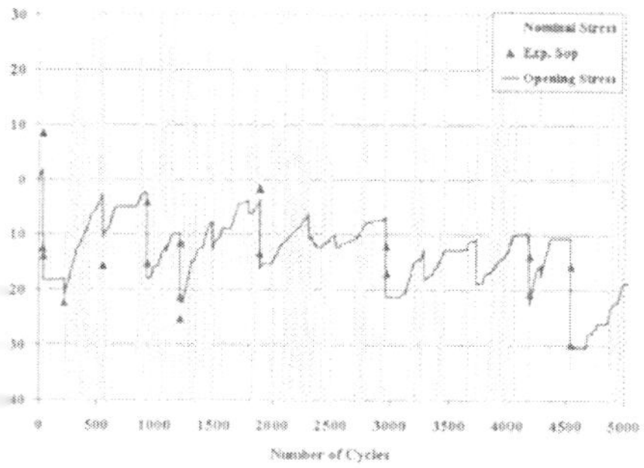

Figure 16 - *Section C.*

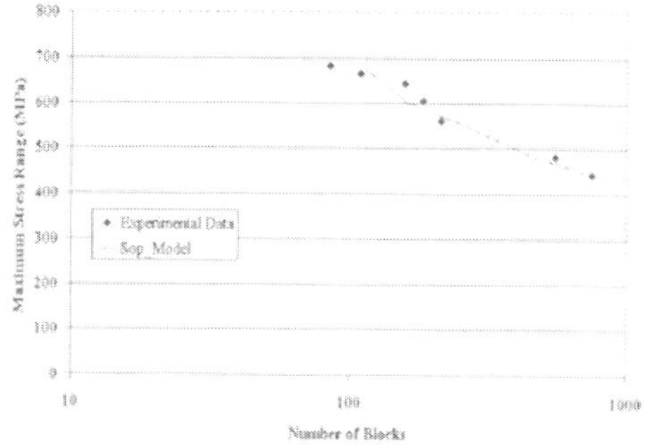

Figure 17 - *Fatigue life vs. maximum stress (IGSK).*

Average Crack Opening Stress

The crack opening stress throughout the fatigue loading did not vary greatly from its average value for each intensity of the load history for the three applied spectra. The number of cycles sharing the same crack opening stress were added together and arranged in ascending order and then added in an accumulative sequence to form the accumulative frequency of occurrence. The accumulative frequency for each cycle is plotted versus its steady state crack opening stress (Figures 18 to 20). Also shown on each curve is the average crack opening stress measured for that applied level of the load history. The frequency of occurrence at which the crack opening stress is equal to the measured average crack opening stress for the history does not vary much with the severity of the load history. An average value of crack opening stress of -16.8, -28, -13

MPa for the severest stress on the three applied spectra respectively gives conservative average closure estimates for the other levels of load history.

Crack closure levels corresponding to these average levels and to the maximum closure in the histories were implemented in the fatigue notch model and fatigue lives were estimated for AL 2024-T351 under the Grapple-Skidder spectrum, the Log Skidder spectrum and the inverse of the Grapple-Skidder spectrum. The results are shown in Figures 21 to 23. The results of the calculations show clearly that fixing the crack opening stress level at the minimum level in the history leads to conservative fatigue life estimations. A good estimate of fatigue life occurs using the chosen S_{op} for a frequency of occurrence of 1/200 for all the histories.

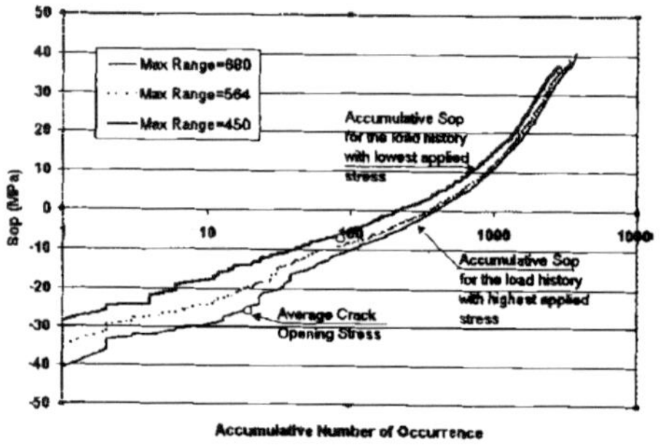

Figure 18 - *Accumulative frequency versus Sop for (GSK).*

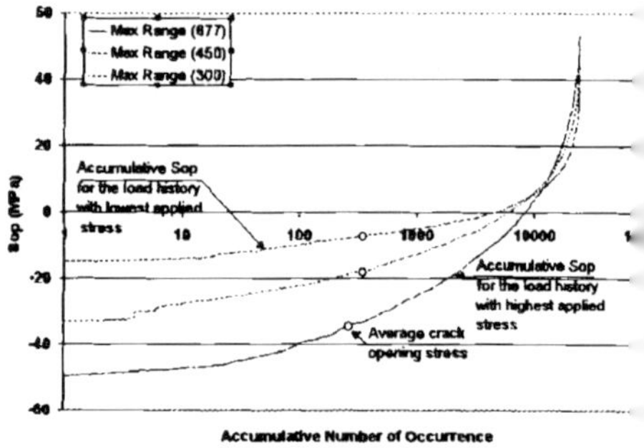

Figure 19 - *Accumulative frequency versus Sop for (LSK).*

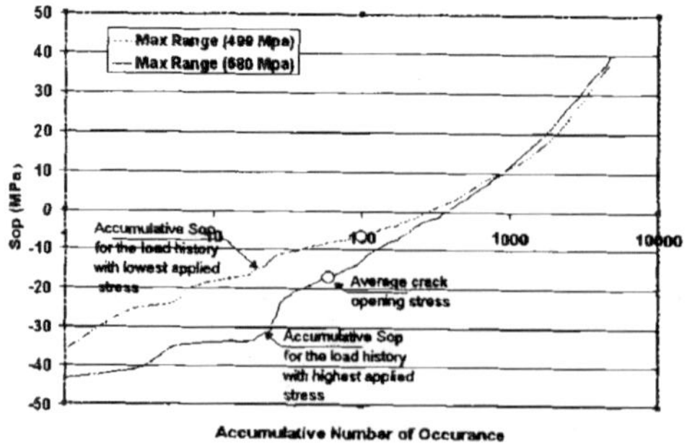

Figure 20 - *Accumulative frequency versus Sop for (IGSK).*

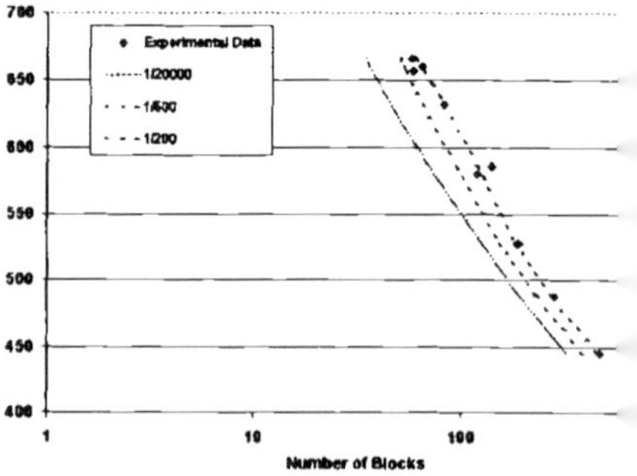

Figure 21 - *Fatigue life versus stress range for (GSK).*

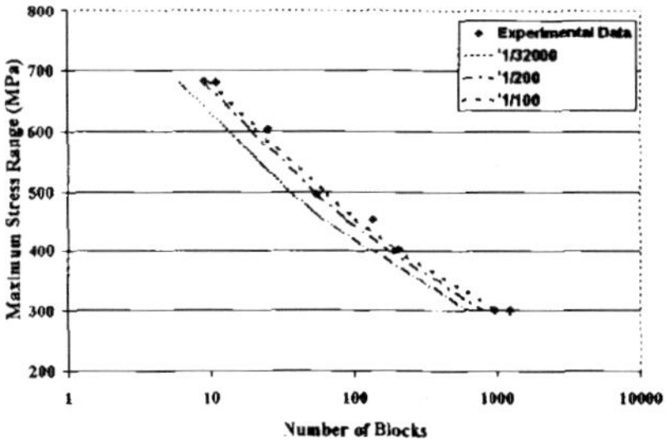

Figure 22 - *Fatigue life versus stress range for (LSK).*

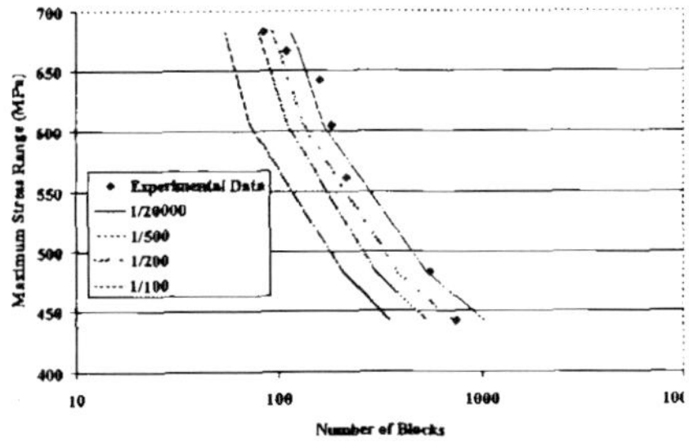

Figure 23 - *Fatigue life versus stress range for (IGSK).*

Conclusions

1. The COS level dropped immediately after the application of an overload and then gradually increased with subsequently small cycles.
2. The crack opening stress can be modeled using an exponential build-up formula which is a function of the difference between the current crack opening stress and the steady state crack opening stress of the given cycle.
3. The measured COS showed that the crack opening stress does not vary greatly from its average value for each intensity of the load history.
4. The fatigue-life of a 2024-T351 component can be modeled using the COS level corresponding to the one in 200 cycle which gave a conservative estimate of average COS for all the histories.

References

[1] Elber, W.," The Significance of Fatigue Crack Closure," *In Damage Tolerance in Aircraft Structures, ASTM STP 486*, American Society for Testing and Materials, Philadelphia, 1971, pp. 230 – 242.

[2] DuQuesnay, D. L.," Fatigue Damage Accumulation in Metals Subjected to High Mean Stress and Overload Cycles," Ph.D. Thesis, University of Waterloo, Waterloo, ON, Canada, 1991.

[3] Bernard, P. J., Lindley T. C. and Richards, C. E., " Mechanics of Overload Retardation During Fatigue Crack Propagation," *ASTM STP 595,* American Society for Testing and Materials, Philadelphia, 1976, pp. 78 - 97.

[4] Wheeler, O. E., " Spectrum Loading and Crack Growth," *Journal of Basic Engineering, Transaction of A SME*, Vol. 4, 1972, pp. 181 - 186.

[5] Matsouks, S. and Tanaka, K., "Delayed Retardation Phenomena of Fatigue Crack Growth in Various Steels and Alloys ," *Journal of Material Science*, Vol. 13, 1978, pp. 1335 - 1353.

[6] Shin, C. S. and Fleck, N. A., "Overload Retardation in a Structural Steel," *Fatigue and Fracture of Engineering Materials & Structure*, Vol. 9, 1987, pp. 379 - 393.

[7] Ward-Close, C. M. and Blom, A. F. and Ritchie, R. O., " Mechanisms Associated with Transient Fatigue Crack Growth Under Variable Amplitude Loading," *Engineering Fracture Mechanics*, Vol. 32, 1989, pp. 613 - 638.

[8] Suresh, S., "Micromechanisms of Fatigue Crack Retardation Following Overloads," *ASTM STP 595*, American Society for Testing and Materials, Philadelphia, Vol. 18, 1983, pp. 577 - 593.

[9] Schijve, J., " The Accumulation of Fatigue Damage in Aircraft materials", *AGARD*, No. 157, 1972.

[10] Topper, T. H. and Yu, M. T., " The Effect of Overloads on the Threshold and Crack Closure," *International Journal of Fatigue*, Vol. 7(3), 1985, pp. 159 - 164.

[11] Alzeo, W. X., Skatjr, A. C. and Hillberry, B. M., " Effect of Single Overload/Underload Cycles on Fatigue Crack Propagation," *ASTM STP 595*, American Society for Testing and Materials, Philadelphia, 1976, pp. 41 - 60.

[12] Pompetzki, M.A., Saper, R.A. and Topper, T.H., " Effect of Compressive Underloads and Tensile Overloads on Fatigue Damage Accumulation in SAE 1045 Steel," *International Journal of Fatigue*, Vol. 12, 1990, pp. 207-213.

[13] Pompetzki, M.A., T.H. Topper, D.L. DuQuesnay and M.T. Yu, "Effect of Compressive Underloads and Tensile Overloads on Fatigue Damage Accumulation in 2024-T351 Aluminum," *Journal of Testing and Evaluation*, Vol. 18, No. 1, 1990, pp. 53-61.

[14] Topper, T. H., M. T. Yu and D. L. DuQuesnay, " Mechanisms and Mechanics of Fatigue Crack Initiation and Growth," *Proceeding Of the International Symposium on Fracture Mechanics*, W. R. Tyson and B. Mukherjee Eds, 1987, pp. 81-94.

[15] Yu, M. T., Topper, T. H., DuQuesnay, D. L. and Levin, M. S., "The Effect of Compressive Peak Stress on Fatigue Behavior," *International Journal of Fatigue*, Vol. 8, 1986, pp. 9 - 15.

[16] Dabayeh, A. And T.H. Topper, " Changes in Crack-Opening Stress after Underloads and Overloads in a 2024-T351 Aluminum Alloy," *International Journal of Fatigue*, Vol. 17, No. 4, 1995, pp. 261-269.

[17] Pompetzki, M.A., Saper, R. A. and T. H. Topper, *Canadian Metal*, Vol. 25, 1986, pp. 181.

[18] Vormwald, M. and T. Seeger, "The Consequences of Short Crack Closure on Fatigue Crack Growth Under Variable Amplitude Loading," *Fatigue and Fracture of Engineering Materials & Structure*, Vol. 14, No. 3, 1991, pp. 205 – 225.

[19] Dabayeh, A. A., Xu, R. X., Du, B. P., and Topper, T. H., "Fatigue of Cast Aluminum Alloys under Constant and Variable-amplitude Loading," *International Journal of Fatigue*, Vol. 18, No. 2, 1996, pp. 95 – 104.

David Alan Johnson[1]

Automated Deformation Mapping in Fatigue and Fracture

Reference: Johnson, D. A., "**Automated Deformation Mapping in Fatigue and Fracture,**" *Applications of Automation Technology in Fatigue and Fracture Testing and Analysis: Fourth Volume, ASTM STP 1411,* A. A. Braun, P. C. McKeighan, A. M. Nicolson, and R. D. Lohr, Eds., American Society for Testing and Materials, West Conshohocken, PA, 2002.

Abstract: An example of the use of deformation mapping in the study of fatigue-crack growth (FCG) has been developed, and preliminary steps in automating this method have been made. The deformation-mapping technique used was based on the correlation of corresponding subsets of digital images acquired while the specimen was under varying loading conditions. The result was a full-field in-plane deformation map which showed, in general, nonlinear behavior under loading. The deformation map of the region near the crack tip was used in calculating both Mode-1 (opening) and Mode-2 (shearing) near-tip stress-intensity-factors. Also, measurements of crack opening displacements (CODs) and crack sliding displacements (CSDs) have been made for various measurement locations. In this way, the effect of measurement location upon closure measurements can be studied. Lastly, a Synthesis of Modeling and Application in Real-time (SMART) system has been proposed, and work on this system has been ongoing.

Keywords: fatigue-crack growth, mixed mode, deformation mapping, image analysis, closure, crack opening displacement, crack sliding displacement

Introduction

Deformation mapping can be of great utility in the study of the mechanical behavior of solids. Results from in-plane full-field deformation mapping can be used to calibrate models and can give useful information concerning material behavior well beyond the capabilities of one-dimensional methods such as strain gages and laser interferometric displacement gages.

Deformation mapping has been applied by many researchers in the past [*1-4*], but these systems have been physically removed from the actual test system and have operated on images recorded on film, although in some cases the film photographs have been digitized. In the present study, a system was developed that is attached to a standard servo-hydraulic test frame and that uses a high-resolution digital camera to acquire and

[1] Materials Research Engineer, Materials and Manufacturing Directorate, Air Force Research Laboratory (AFRL/MLLMN), Building 655, Suite 1, 2230 Tenth Street, Wright-Patterson Air Force Base, OH, USA 45433-7817.

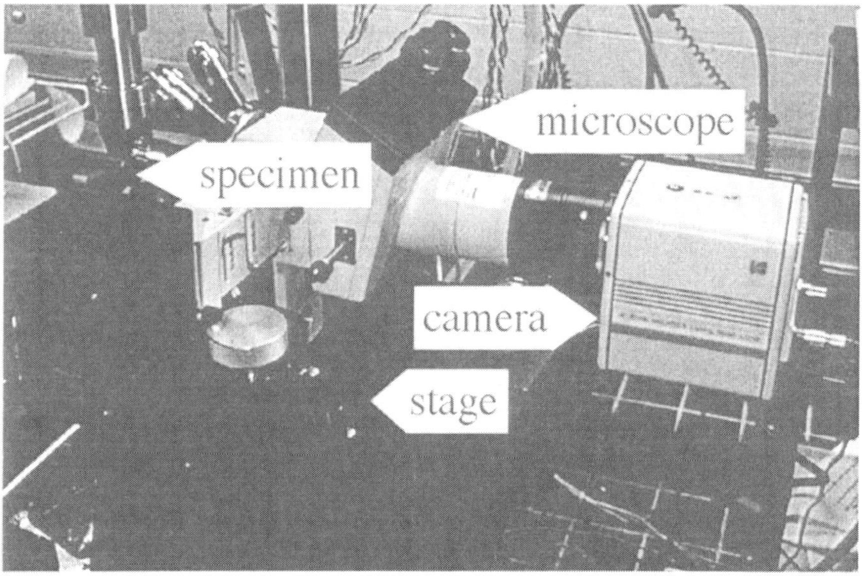

Figure 1 – *Deformation-Mapping Equipment.*

process images during testing. In this way, adjustments to the test based on the calculations from the deformation maps can eventually be made.

Experimental System and Deformation-Mapping Process

Figure 1 illustrates the experimental equipment for deformation mapping with the test specimen, digital camera, manual stage, and microscope labeled. The test specimen is mounted in a standard servo-hydraulic loading frame of the type typically used in fracture and fatigue tests. This particular specimen is of a Brazilian-disk geometry, but any specimen having a flat surface in the region of interest and having little or no out-of-plane displacement can be used. Out-of-plane displacements can result in false deformations being included in the deformation-mapping process.

The digital camera has high spatial resolution (nominally 2000 by 2000 pixels) and is controlled by a computer through the use of virtual-instrument-based software. This same software is capable of fully controlling the loading of the specimen and all other aspects of the test. The microscope is a standard type used in materials science with five different objective lenses and five different photo eyepieces, giving a very wide selection of magnifications. The microscope and camera are mounted on a three-degree-of-freedom manually operated stage which is attached to the load frame. In the future, an automated stage, controllable by the same virtual-instrument-based software, will be used, greatly enhancing the ability of the system to be automated.

The initial step in the deformation-mapping process is to acquire at least two images for analysis. These images are taken of the same region of a solid under differing

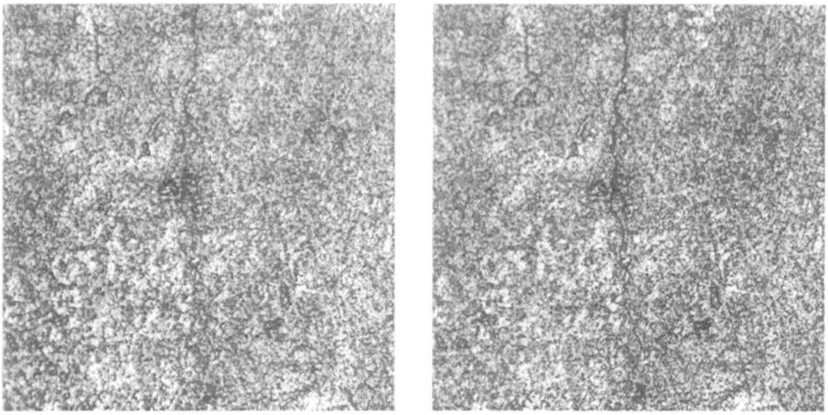

Figure 2 – *Examples of Undeformed (left) and Deformed Images.*

conditions. The change in conditions is typically mechanical loading, although changes in temperature or any other process producing deformation could be studied. For illustration, the following will discuss only two images, one acquired under "unloaded" and the other under "loaded" conditions. However, the concepts are easily extended to a sequence of images, taken under increasing or decreasing loading or both, and this has been done, enabling the study of nonlinear time-dependent or non-time-dependent behavior.

The calculation of the relative deformation between the unloaded and loaded images is accomplished using a subset-correlation technique. The technique used is an adaptation of a reasonably well-known technique, so only the basic outline of the technique will be described here. In this process, the images under unloaded conditions and loaded conditions, referred to as the undeformed and deformed images, respectively, are taken of the same area of the solid under consideration, within a small error. Standard two-dimensional correlation techniques [*5*] are used to align the camera so as to maximize the correlation between the deformed and deformed images. Also, numerical techniques are used to minimize the effect of large-scale light-intensity gradients and to normalize the distribution of the intensities [7]. Figure 2 shows examples of an undeformed image and the corresponding deformed image after this processing.

To determine the deformation map from the images, the same correlation technique is applied to square subsets of the images. The size of the square subsets can be varied based on the quality of the images and the desired information. Large subsets assure good correlation but may give poor resolution of local deformation. Small subsets generally yield good resolution of local deformations but can result in poor correlation. Also, the subsets are generally overlapped to give more values of deformation.

Figure 3 shows the undeformed image with the overlapping squares used for calculating the deformation vectors. One particular square is highlighted, and for the square the objective is to determine the displaced positions in the deformed image corresponding to the points of the solid in this square. This is accomplished by using correlation techniques on the corresponding area of the deformed image, which is also

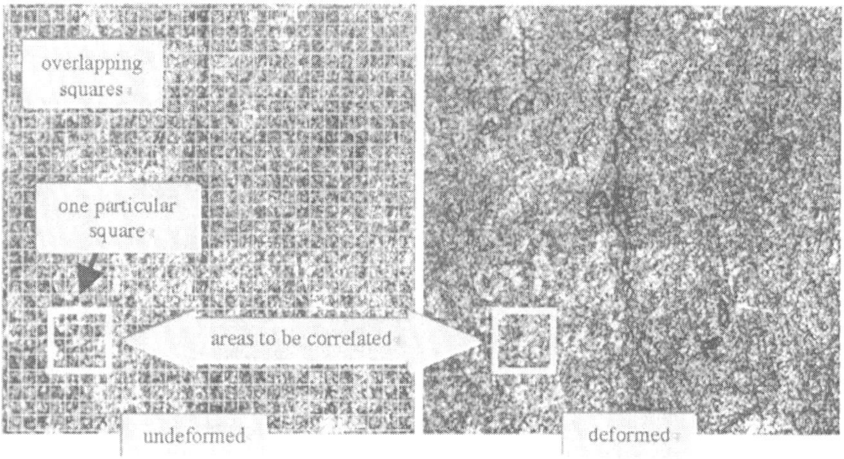

Figure 3 – *Schematic of the Process for Determining Deformation Vectors.*

highlighted. An implicit assumption being made is that the deformations within the square are small enough that the correlation can be done well. In the case of high-strain materials, such as rubbers, this assumption may be incorrect. In this case, it is better to use a correlation technique that includes the displacement gradients, and not just the displacements, as part of the correlation function, as has been done [*1*]. For the materials under study here, inclusion of the displacement gradients in the correlation process is not necessary and may result in computational inefficiency. The primary reason for this inefficiency is that fast-Fourier-transform techniques cannot be used if the displacement gradients are included.

Figure 4 shows corresponding squares (64 X 64 pixels) from the undeformed and deformed images, along with their associated correlation matrix, both as a density graph and a surface graph. The maximum of the correlation matrix is obvious, and the coordinates of this point represent the coordinates of the desired displacement vector for this particular square. However, the element indices will only give the displacements to within one pixel value. One group of researchers [*3*] used a two-dimensional second-order fit of the correlation matrix to estimate the displacement to better than one pixel. In this study, a full bicubic interpolation [*5*] has been done of the entire region near the correlation-matrix maximum, which gives much better than one-pixel resolution. Another method for increasing the resolution of the technique is to perform a bicubic interpolation on each of the entire images (or all of the images, if there is a sequence of images) before processing, at some cost in computational efficiency and storage capacity. Too much interpolation (generally more than twice the original image size) will result in no improvement, however, because there is no new information being generated in the interpolation process.

The technique just described is performed on all of the overlapping squares, resulting in a set of two-dimensional vectors representing the overall displacements of the squares. In many cases, there is a relatively large amount of "rigid-body motion," both translation and rotation, between the undeformed and deformed images. This rigid-body motion

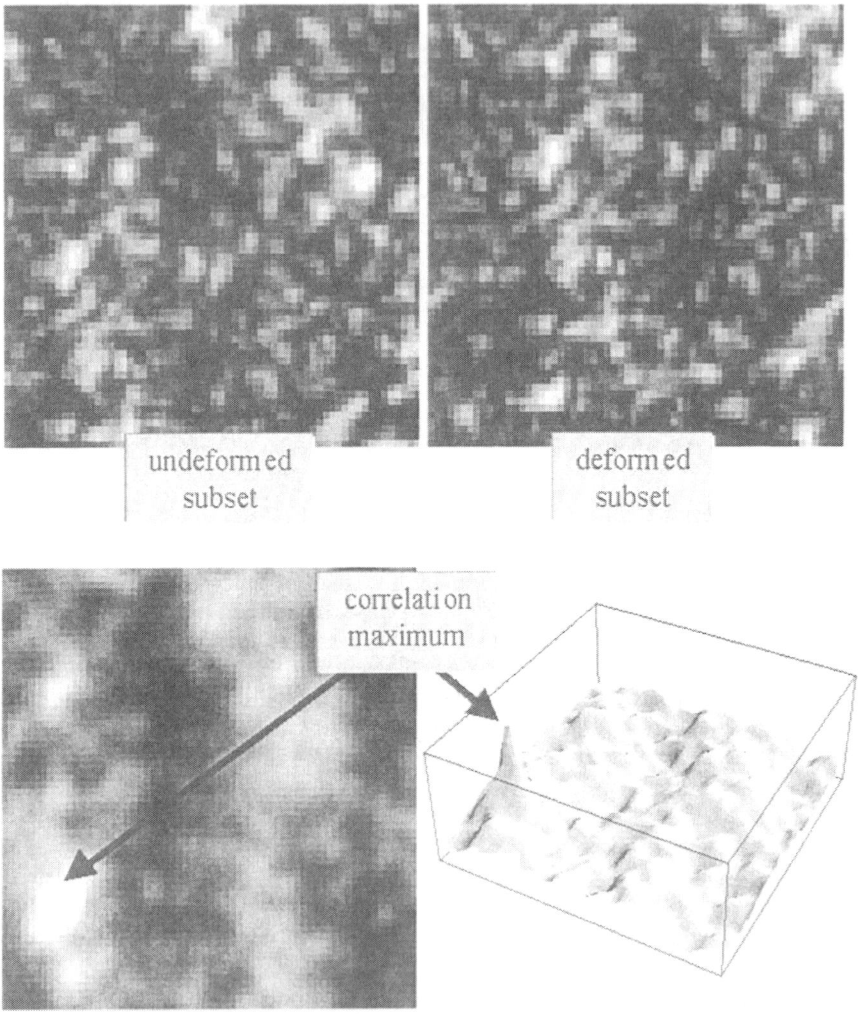

Figure 4 – *Square Subsets for Determination of One Displacement Vector (top) and Density and Surface Plots of their Correlation Matrix (bottom).*

may or may not correspond to actual rigid-body motion of the solid under study. It may just be relative motion of the region of interest (ROI) in a complex geometry and not overall motion. Whatever the cause, it can interfere with subsequent analyses, especially for a sequence of images acquired to study nonlinear behavior, and a decision must be made on how to minimize the effects of this motion. One possible approach, and the one used here, is to consider a point in the ROI to be a reference point, and to subtract the calculated displacement vector of this point from every vector, forcing that point's rigid-

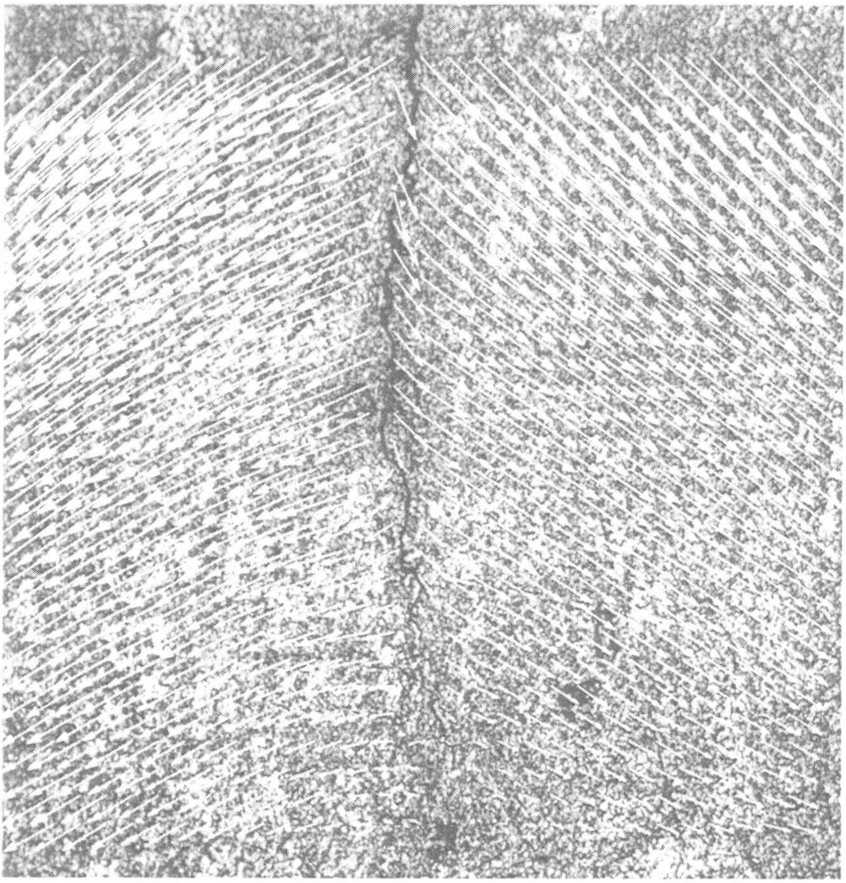

Figure 5 – *Deformation Map with Field of View of 0.6 mm.*

body translation to be zero. Then a rotation about the reference point is calculated, using optimization techniques, which will minimize the sum of the magnitudes of all of the vectors. Many times, this reference point will be the tip of a fatigue crack or some other point of interest in the image.

Figure 5 shows the result of the deformation mapping for the two example images. For clarity, the number of vectors in the figure is much less than the number that would normally be calculated, and the vectors have been exaggerated 50 times relative to the image scale. The relative resolution of the deformation-mapping technique is not directly related to the size of the area under study but is instead dependent solely on the quality of the initial images. However, with an optical system as used here the image quality will generally degrade with increased magnification, resulting in a lower-quality deformation map. Also, as mentioned before, the size of the subset squares used to calculate the displacement vectors is of importance, with smaller squares giving better representation of localized displacements but possibly with worse correlation, resulting in more scatter in the deformation-map data. Another important point is that the surface being imaged must be highly featured at the scale of interest since the correlation is done on the intensity differences resulting from the features.

Calculation of Surface-Deformation Stress-Intensity Factors (SDSIFs) from Deformation Maps

One use of deformation mapping particularly relevant to fracture and fatigue is the calculation of SDSIFs from deformation maps. Both Mode-1 (opening) and Mode-2 (shearing) SDSIFs can be calculated from in-plane deformation maps. These SDSIFs have been designated K_1^{sd} and K_2^{sd}, respectively. Comparison of these SDSIFs with the stress-intensity factors (SIFs) calculated from the global loading and geometry can yield valuable insight into the effects of crack closure and other nonlinear phenomena.

Figure 6 shows typical surface plots of the horizontal displacement, *u*, and vertical displacement, *v*, near a crack tip, with the crack path generally in the horizontal direction, along with the corresponding vector map superimposed on the deformed image. Once again, the displacement vectors have been exaggerated for illustration. For each displacement plot, the units on the two horizontal axes, representing the spatial coordinates, and the vertical axis, representing the displacement value for that particular coordinate, are all in μm.

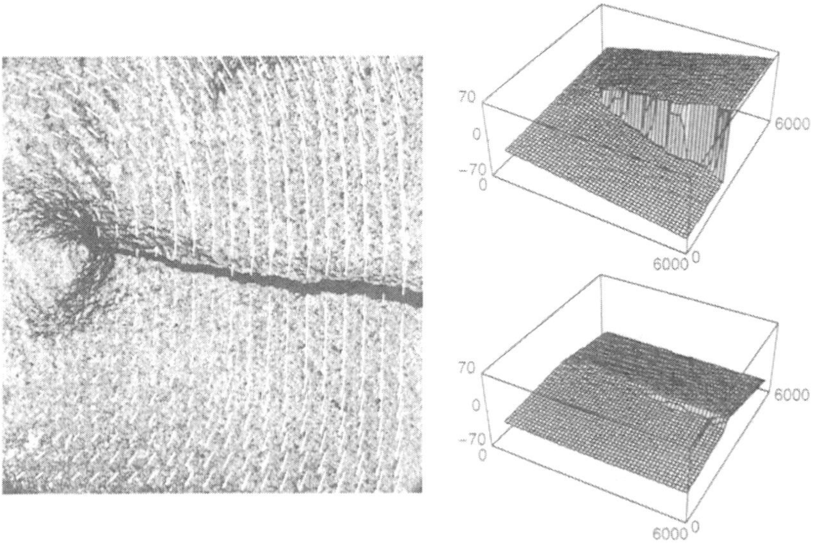

Figure 6 – *Vector Map Superimposed on Deformed Image (left) and Displacement in the Horizontal Direction (bottom right) and Vertical Direction (top right) near the Tip of a Crack Growing Generally Horizontally. All Values in μm.*

Calculating K_I^{sd} and K_2^{sd} is accomplished by fitting the results of a calculation of displacements from classic linear-elastic fracture mechanics (LEFM) near a crack tip [*6*] to the results obtained from the deformation-mapping process. The parameters of the model include the angle of the crack and the crack-tip location, along with the usual material and LEFM parameters. It is recognized that as one moves away from the crack tip the displacements are influenced by geometric considerations and should not match exactly the displacements calculated using the crack-tip displacement-field equations. However, it can be shown [*7*] that the differences between the displacements of a finite-width specimen and the crack-tip displacement-field results are relatively small to a reasonably far distance from the crack tip. Also, in the future the models used will be much more sophisticated than the simple LEFM model used here. These models can be elastic or elastic/plastic and can be either entirely analytical (as here) or from finite-element analysis results.

The crack shown in Figure 6 was in a compact tension, C(T), specimen with a width, W, of 39.89 mm and a thickness, B, of 2.15 mm, as defined by the ASTM Standard Test Method for Measurement of Fatigue Crack Growth Rates (E 647). The surface crack length, a, was determined through optical measurement, again according to E 647, to be 33.22 mm (with one side as 33.20 mm and the other as 33.23 mm). With the load applied as 650 N, this should result in a Mode-1 stress-intensity factor, K_I , of 82.8 MPa$\sqrt{m}$. However, it should be noted that this particular combination of crack length, geometry, and loading violates the requirement of E 647 that the specimen be predominantly elastic (section 7.2.1 of E 647). This explains in large part the discrepancy between K_I^{sd} and K_I noted below. Another problem is that at these relatively high values of $\frac{a}{W}$ in the C(T) specimen, a small error in measured crack length or the difference between the surface-measured crack length and the curvature-corrected crack length can create large changes in K_I. Work is ongoing in studying the correlation of the stress-intensity factors calculated from geometry and the surface-measured stress-intensity factors under various conditions. Normally, the stress-intensity factors under consideration and the displacements near the crack tip are much smaller than in this example, which was chosen for illustration.

Figure 7 shows the results of the LEFM model next to the measured displacements. The actual fitting is accomplished with a minimization scheme where the sum of the squares of the differences between the measured and model displacements is the function to be minimized. Details may be found in Reference 7. As can be readily seen, the fit is quite good, although the calculated values of K_I^{sd} and K_2^{sd} (119 and −1.3 MPa$\sqrt{m}$) are different than K_I and K_2 as determined strictly through geometry and loading (82.8 and 0.0 MPa$\sqrt{m}$). The probable reasons for this are cited above. However, even this result is of interest in the study of the effect of the decreasing width of remaining specimen ligament on crack-tip behavior.

Results for a nominally pure-Mode-1 specimen geometry, the C(T), have been shown, but in-plane mixed-mode specimens can also be easily studied and K_I^{sd} and K_2^{sd} can both be determined as shown. In this way, an attempt can be made to correlate crack-growth rate with Mode-1 and Mode-2 stress-intensity factors. This is of special interest in materials where self-similar mixed-mode crack growth is common, such as single-crystal superalloys. Details of some mixed-mode results may be found in Reference 7.

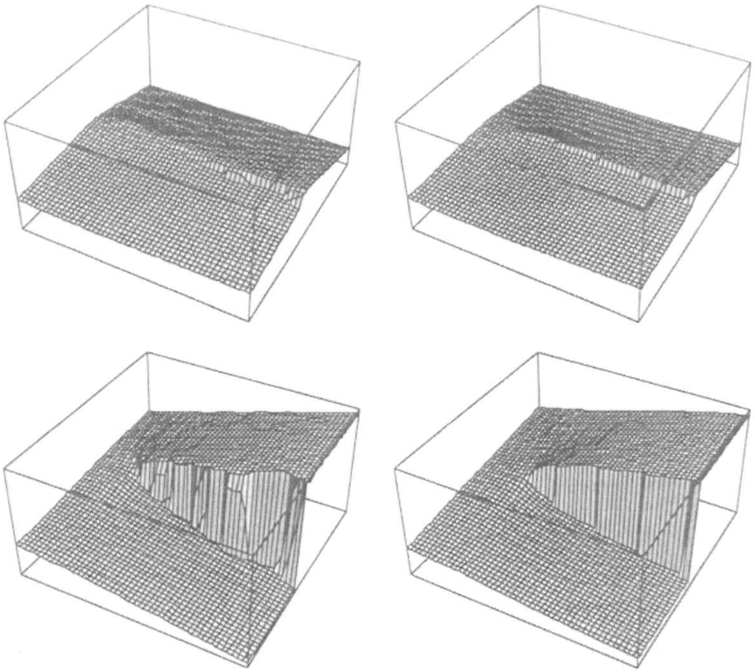

Figure 7 – *Measured Displacement in the Horizontal Direction (top left) and Vertical Direction (bottom left) and Corresponding Model Displacements (right). Scales same as Figure 6.*

Study of Nonlinear Behavior

By acquiring images at various load levels for a solid of constant geometry, it is possible to study nonlinear behavior. For fatigue-crack growth, a nonlinear behavior of great interest is crack closure. To determine crack closure for a particular crack length, the process as described above is performed on the images acquired at each load level. In each case, the image at that particular load level is the deformed image as defined above, and the image at no load of at the minimum load is defined as the undeformed image. It is critical that the processes applied to each of the images, such as intensity smoothing and rigid-body-displacement correction, are consistent for all images.

Figure 8 shows the results of the calculation of K_I^{sd} and K_2^{sd} for many load levels. The stress-intensity factor corresponding to the maximum load for this plot is much less than that for Figure 6, and it should be much more appropriate to apply LEFM to this case. As can be seen, the closure, which might be called "full-field closure," is quite evident. By use of this technique, it is possible to study closure in both Mode 1 and Mode 2 as has been done with a point-measurement technique [*8*], but which can now be done with a full-field technique.

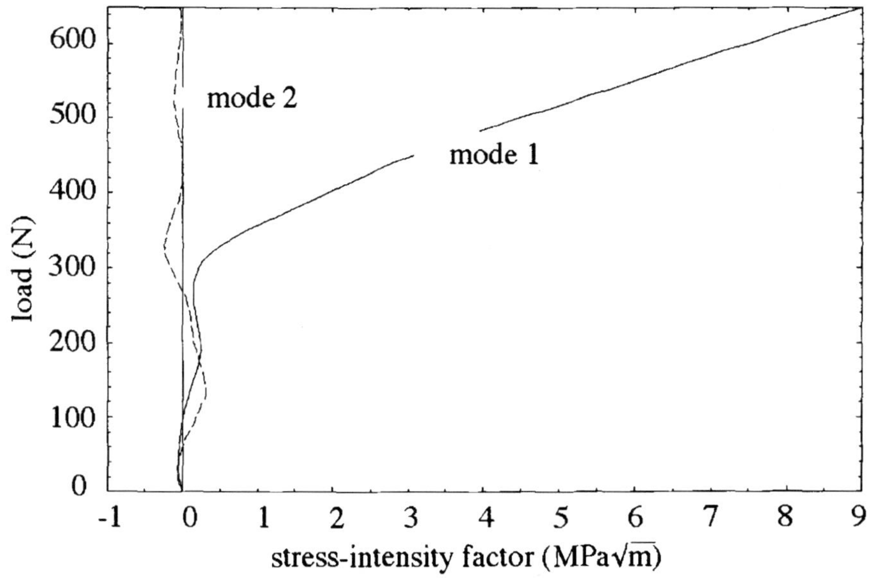

Figure 8 – *Full-Field Closure for both Mode 1 and Mode 2.*

Another aspect of closure that can be studied is the effect of measurement location on closure determination. It is quite easy to arbitrarily pick a distance from the crack tip for measuring crack opening displacement (COD) and crack sliding displacement (CSD). Figure 9 shows the behavior of the load/COD relationship as measurement location is varied. These data result from the mosaicking of several images, as is shown in the figure. As expected, the COD increases with distance from the crack tip, but it is also evident that the "knee" in the load/COD behavior is more pronounced and the apparent closure level increases as the crack tip is approached.

Discussion

Although this paper has concentrated on the use of deformation mapping in fatigue-crack growth, deformation mapping may be applied in any field which involves the deformation of solid materials. For example, in-plane material parameters, such as E_{11}, E_{22}, ν_{12}, and so on, can be easily determined through the study of the deformation of a specimen undergoing a tension test. By continuing the test into the inelastic regime, it is possible to study the inelastic behavior of solids. Also, by the use of two or more imaging systems, it is possible to calculate three-dimensional material parameters in both the elastic and inelastic regimes.

In the area of fatigue, changes in materials undergoing fatigue cycles can be studied, along with the effect of surface treatments on both fatigue and small-crack fatigue-crack growth. If initiation sites can be determined, the process of fatigue-crack initiation, including the effect of microstructure, can also be studied.

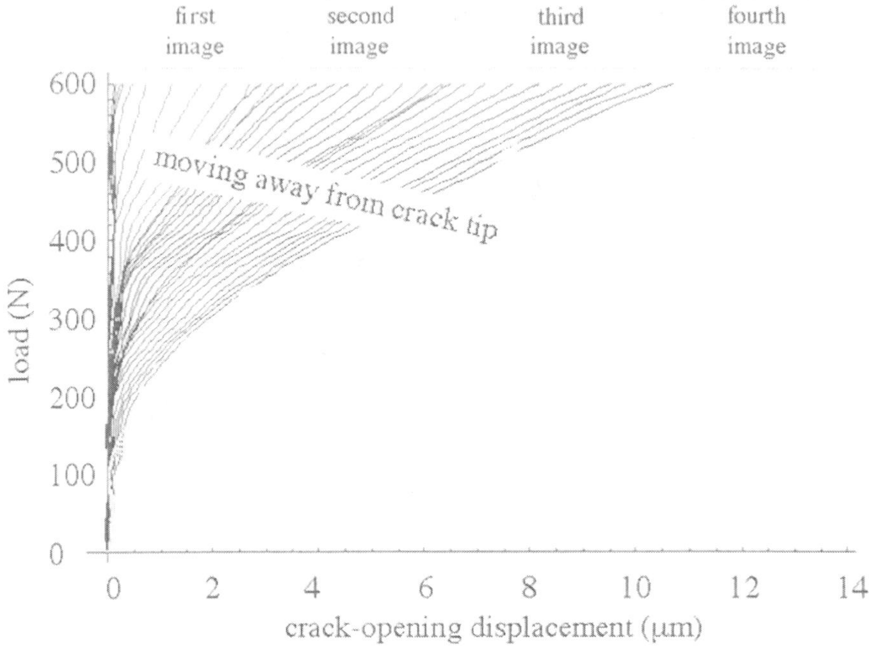

Figure 9 – *Effect of Measurement Location on Closure Determination.*

In order to generalize and facilitate the application of deformation mapping, the Synthesis of Modeling and Application in Real Time (SMART) system is under development. As shown schematically in Figure 10, this system both studies a solid body with an imaging system and simultaneously models that body. By correlating the deformation predicted by the model with the actual deformation, the model can be calibrated and evaluated in quasi-real time. The model can be analytical, numerical, or a combination of both. It can, in general, be time, temperature, and load-dependent. The microstructure and macrostructure of the body can be both determined through imaging and modeled. The body can be a material undergoing processing or a specimen or component undergoing test.

The biggest challenges for the SMART system are the speed at which deformation mapping can be accomplished, the automation of the image-acquisition process, and the ability to do deformation mapping at use temperatures well above room temperature. The computational-speed problem should be alleviated as computers grow in power, but the problems with acquiring undistorted magnified images at elevated temperatures are immense. However, these problems are tractable with sufficient expertise. Some work has been done on unmagnified images at elevated temperatures [9], although the addition of magnification ability adds immensely to the difficulties in obtaining good quality images at elevated temperatures. The automation of the imaging system can be

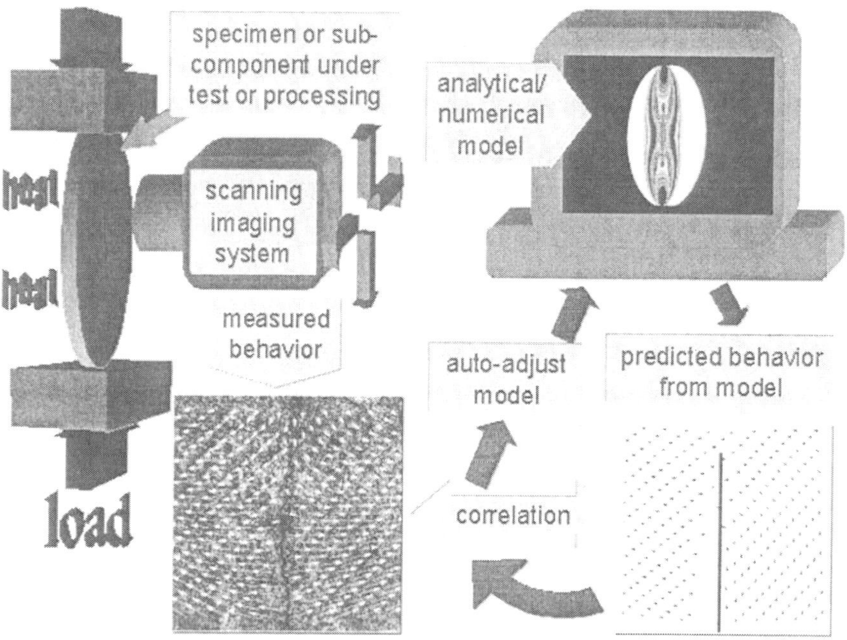

Figure 10 – *Schematic of the Synthesis of Modeling and Application in Real Time (SMART) System.*

accomplished by the installation of an automated stage in place of the manual stage shown in Figure 1. Sophisticated software is available to control the stage so as to acquire the images with precise spatial precision. Through automation, large areas of the solid can be studied without human intervention.

Conclusion

The future of deformation mapping in the fields of fracture and fatigue offers much promise, especially after necessary steps in the automation of the image-acquisition process are accomplished. The utility of deformation mapping in the study of crack opening and crack sliding displacements (CODs and CSDs) has been demonstrated, along with the possibilities of the closer integration of modeling and experimentation as shown by the Synthesis of Modeling and Application in Real Time (SMART) system. The largest challenges to the application of automated deformation mapping to fracture and fatigue are the speed at which the highly complex calculations can be made and the difficulty of obtaining high-quality images at use temperatures which can be well above room temperature.

Acknowledgments

This work was supported by the United States Air Force Research Laboratory, Materials and Manufacturing Directorate, Metals, Ceramics, and Nondestructive Evaluation Division, Metals Branch (AFRL/MLLMN) and the Air Force Office of Scientific Research, and that support is gratefully acknowledged. In addition, the author wishes to acknowledge his invaluable discussions with Drs. J. M. Larsen and R. John of the AFRL, Dr. D. L. Davidson of the Southwest Research Institute, Prof. F.-P. Chiang of the State University of New York at Stony Brook, and the author's thesis advisor, Prof. J. W. Hutchinson of Harvard University.

References

[1] Bruck, H., McNeill, S. R., Sutton, M. A., "Digital Image Correlation Using the Newton-Raphson Method of Partial Differential Correlation," *Experimental Mechanics*, Vol. 29, No. 3, September, 1989, pp. 261-267.

[2] Franke, E. A., Wenzel, D. J., and Davidson, D. L., "Measurement of Microdisplacements by Machine Vision Photogrammetry (DISMAP)," *Review of Scientific Instruments*, Vol. 62, No. 5, May, 1991, pp. 1270-1279.

[3] James, M. R., Morris, W. L., and Cox, B. N., "A High Accuracy Automated Strain-field Mapper," *Experimental Mechanics*, Vol. 30, No. 1, March, 1990, pp. 60-67.

[4] Chen, D. J., Chiang, F. P., Tan, Y. S., and Don, H. S., "Digital Speckle-Displacement Measurement Using a Complex Spectrum Method," *Applied Optics*, Vol. 32, No. 11, April, 1993, pp. 1839-1849.

[5] Press, W. H., Teukolsky, S. A. Vetterling, W. T., and Flannery, B. P., *Numerical Recipes in C: The Art of Scientific Computing, Second Edition*, Cambridge University Press, Cambridge, England, 1992, pp. 125-128, 545-546.

[6] Broek, D., *Elementary Engineering Fracture Mechanics, Fourth Revised Edition*, Martinus Nijhoff Publishers, Dordrecht, The Netherlands, 1986, pp 77-81.

[7] Johnson, D. A., "Mixed-Mode Fatigue-Crack Growth in Turbine-Engine-Rotor Materials," Ph.D. Thesis, Harvard University, Cambridge, Massachusetts, to be published, 2001.

[8] Campbell, J. P., and Ritchie, R. O., "Mixed-Mode, High-Cycle Fatigue-Crack Growth Thresholds in Ti-6Al-4V: II. Quantification of Crack-Tip Shielding," *Engineering Fracture Mechanics*, Vol. 67, No. 3, October, 2000, pp. 229-249.

[9] Lyons, J. S., Liu, J., and Sutton, M. A., "High-Temperature Deformation Measurements Using Digital-Image Correlation," *Experimental Mechanics*, Vol. 36, No. 1, March, 1996, pp. 64-70.

Steven J. Gill[1] and Peter S. Pao[2]

A Method for Conducting Automated Fatigue Crack Initiation Tests on Fracture Mechanics Specimens

Reference: Gill, S. J. and Pao, P. S., "**A Method for Conducting Automated Fatigue Crack Initiation Tests on Fracture Mechanics Specimens,**" *Applications of Automation Technology in Fatigue and Fracture Testing and Analysis: Fourth Volume, ASTM STP 1411*, A. A. Braun, P. C. McKeighan, A. M. Nicolson, and R. D. Lohr, Eds., American Society for Testing and Materials, West Conshohocken, PA, 2002.

Abstract: Fatigue crack initiation in notched members is controlled by local strains at the notch root. A number of approaches have been developed for calculating local notch-tip stresses and strains from nominal stress and notch geometry considerations. One such approach uses the parameter $\Delta K / \rho^{1/2}$, where ΔK is the fracture mechanics stress intensity range and ρ is the notch root radius. The parameter $\Delta K / \rho^{1/2}$ has been shown to correlate with local notch-tip strain and provide a means of normalizing cycles-to-initiation, N_i, data for various notch-tip geometries. Fatigue crack growth rate specimens described in ASTM Test Method for Plane-Strain Fracture Toughness of Metallic Materials (E 399), ASTM Test Method for Measurement of Fatigue Crack Growth Rates (E 647) and elsewhere can be used for fatigue crack initiation testing if they have blunt notches. Data in the form of parameter $\Delta K / \rho^{1/2}$ versus N_i has the same units as the traditional *S-N* curves. The advantage of having data in the form parameter parameter $\Delta K / \rho^{1/2}$ versus N_i is that information is available on ΔK and ρ. As in all fracture mechanics testing, data obtained on one specimen geometry can be applied to a wide variety of structural geometries. This approach can be used in conjunction with commercially available software for fatigue crack growth rate testing, servohydraulic testing equipment, and modified fracture mechanics specimens to automate fatigue crack initiation testing. The combination of all of these elements represents a new test method. Results are presented for aluminum alloy 7075, where the effects of corrosion pits were studied and for titanium alloy Ti-6Al-4V, where the effects of heat treatment on initiation were studied.

Keywords: constant amplitude, environment-assisted cracking, fatigue, fatigue crack initiation, fracture mechanics, initiation, metallic materials, stress intensity range

[1]Special Assistant, Materials Science and Component Technology Directorate, Naval Research Laboratory, 4555 Overlook Ave., SW, Washington, DC, 20375-5343.
[2]Materials Research Engineer, Physical Metallurgy Branch, Materials Science Division, Naval Research Laboratory, 4555 Overlook Ave., SW, Washington, DC, 20375-5343.

Nomenclature

a	crack depth measured from the load line, mm
a/W	normalized crack depth
B	specimen thickness, mm
da/dN	fatigue crack propagation rate, mm/cycle
ΔK	fracture mechanics stress intensity range, MPa÷m
ΔK_{th}	threshold stress intensity for fatigue crack growth, MPa÷m
$\Delta K / \rho^{1/2}$	stress range at the notch-tip, MPa
ΔP	load range, Newtons
$\Delta \sigma_{max}$	local stress range at the notch-tip, MPa
$\Delta \sigma_n$	nominal stress range at the notch-tip, MPa
E	Young's modulus, GPa
H	specimen height, mm
K	fracture mechanics stress intensity, MPa÷m
K_I	mode I fracture mechanics stress intensity, MPa÷m
K_{IC}	fracture toughness (maximum stress intensity material can withstand), MPa÷m
K_{max}	maximum stress intensity in a loading cycle, MPa÷m
K_t	theoretical elastic stress concentration factor
N	elapsed load cycles
N_i	load cycles to initiation
ρ	notch root radius, mm
P	load, Newtons
P_{max}	maximum load in the loading cycle, Newtons
R	load ratio (ratio of minimum to maximum load in a load cycle)
S	stress, MPa
σ_{max}	local stress at the notch-tip, MPa
σ_n	nominal stress at the notch-tip, MPa
σ_{uts}	ultimate strength, MPa
σ_{ys}	yield strength, MPa
W	specimen width measured from the load line, mm
$W - a$	remaining ligament of specimen, mm
x	distance from load line to crack mouth opening measurement location, mm

Introduction

The fatigue life of structures is composed of fatigue crack initiation and fatigue crack propagation. At low stress amplitudes, the majority of the life is often spent in the initiation phase. At the opposite extreme, new high strength alloys often have small critical flaw sizes and thus most of the lifetime of structures made from these alloys is spent in initiation. Therefore, initiation is an important component of the lifetime and the factors that affect it are important determinants of total life. Smooth test specimens that do not contain a precrack are useful for investigation of these factors.

Fatigue crack initiation in notched members is controlled by local strains at the notch root. A number of approaches have been developed for calculating local notch-tip stresses and strains from nominal stress and notch geometry considerations. One such approach uses the parameter $\Delta K / \rho^{1/2}$, where ΔK is the fracture mechanics stress intensity range and ρ is the notch root radius. The parameter $\Delta K / \rho^{1/2}$ has been shown to correlate with local notch-tip strain and provide a means of normalizing cycles-to-initiation, N_i, data for various notch-tip geometries. Fatigue crack initiation data from blunt notched fracture mechanics specimens has been shown to correlate well with results obtained from notched axisymmetric specimens with $K_t = 4$ [*1*].

For wedge-opening-loaded (WOL) [*2*] and compact tension specimens the nominal stress range at the notch-tip ($\Delta\sigma_n$) is given by [*3*]

$$\Delta\sigma_n = \Delta P\, [1 + 3\, ((W + a) / (W - a))] / [B\, (W - a)] \tag{1}$$

Where ΔP is the applied load range, B is the specimen thickness, W is the specimen width measured from the load line, and "a" is the crack depth measured from the load line. The maximum applied stress range, $\Delta\sigma_{max}$, was determined from the generalized stress intensity expression that shows the effect of blunt notches on K_I; where [*4*]

$$\Delta K_I = \lim_{\rho \to 0} \frac{\sqrt{\pi}}{2} \sqrt{\rho}\ \Delta\sigma_{max} \tag{2}$$

The relationship between the local stress range at the notch-tip ($\Delta\sigma_{max}$) and $\Delta K / \rho^{1/2}$ is as follows

$$\Delta\sigma_{max} = 2\, \Delta K / \rho^{1/2} \tag{3}$$

The theoretical elastic stress concentration factor (K_t) can be calculated by taking the ratio of σ_{max} / σ_n for a given specimen geometry and load range

$$K_t = \sigma_{max} / \sigma_n \tag{4}$$

Although Equations 2 and 3 are considered exact only when ρ approaches zero and thus, cannot be arbitrarily applied to large radii, Wilson and Gabrielse have shown as the result of a detailed finite element analysis of blunt notches in compact tension specimens that this expression is accurate to within 10 percent for notch radii up to 4.6 mm [*5*].

Data in the form of $\Delta K / \rho^{1/2}$ versus N_i has the same units as the traditional *S-N* curves. The advantage of having data in the form of $\Delta K / \rho^{1/2}$ versus cycles to initiation is that information is available on ΔK and ρ. As in all fracture mechanics testing, data obtained on one specimen geometry can be applied to a wide variety of structural geometries.

Previous investigations of fatigue crack initiation using this type of specimen [*1, 6-19*] have not used an automated data acquisition technique and were not always able to produce data without a large amount of scatter.

Specimen Configuration and Preparation

Fatigue crack growth rate specimens described in ASTM E 399, ASTM E 647 and elsewhere [*2*] can be used for fatigue crack initiation testing if they have blunt notches [*1, 6-19*]. A typical specimen is shown in Fig. 1.

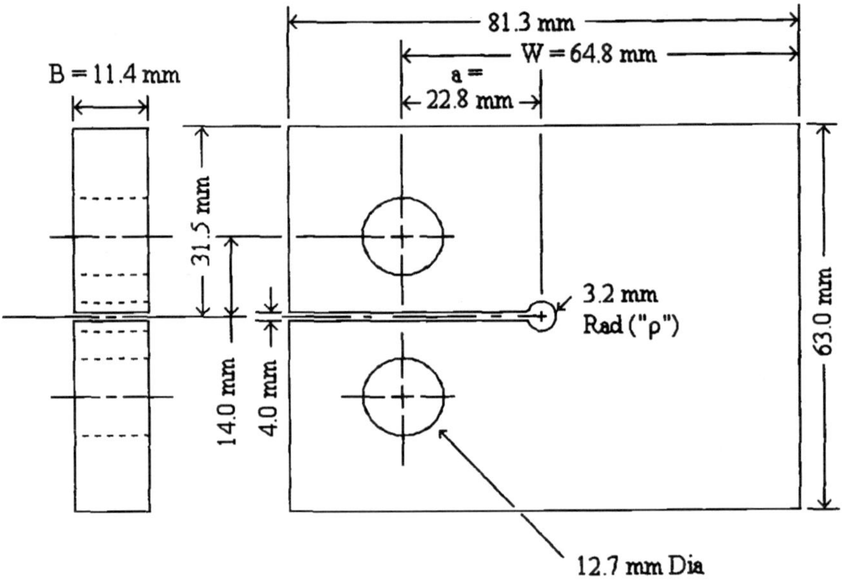

Fig. 1 – *Typical blunt notch fracture mechanics specimen for fatigue crack initiation testing.*

There are limitations on the various parameters, which are as follows

Parameter	Limitation
a	1) $0.2 < a/W < 0.8$ (range of validity of compliance expression) 2) $(a + x) <$ microscope focal length if a microscope is used
B	1) $W/20 < B < W/4$
ΔK	1) $W - a < (4/\pi)(K_{max}/\sigma_{ys})^2$ 2) Must exceed ΔK_{th} 3) $\Delta K/(1 - R) < K_{IC}$

ΔP — P resolution

ρ — Only available drill sizes can be used. This affects *da/dN* after initiation through the $\Delta K / \rho^{1/2}$ parameter.

W — Limited to 4.0 times the pin hole diameters available for compact tension specimens or 5.1 times the pin hole diameters available for WOL specimens.

Steps were taken to ensure that all scratches at the notch root were parallel the direction of applied stress to ensure that the notch root radius would control initiation. If the scratches had been perpendicular to the applied stress then the radius of the scratches, rather than the radius of the notch, would control initiation. This was undesirable, as the radius of the scratches was not easily controlled. The notches were sanded using a rod of a smaller diameter than the notch. The rod was slit lengthwise and used to hold sandpaper turned by a drill press or portable drill. The slot was used so that if sandpaper with an adhesive backing was not available then the sandpaper could be attached to the rod by putting it in the slot and then wrapping it around the rod. The sandpaper was always kept turning when it was in contact with the notch radius to keep the sanding marks in the desired direction. The notches were then polished with metallographic polishing compounds applied to cotton swabs. The swabs were always kept turning when in contact with the notch radius to keep the polishing marks in the desired direction. The final polish was performed with a 3 μm diamond paste. A previous study [*12*] on Ti-6Al-4V titanium alloy specimens showed no significant difference between the results for specimens polished to a 1.83 μm finish and those polished to a 0.41 μm finish. It is probably more important that the polishing marks be perpendicular to the cracks than that the marks be smaller than a certain size. Polishing compounds were removed for inspection of the notch radius by rinsing. The final rinse was with alcohol or acetone to remove any water remaining from the polishing compound or an earlier rinse. If rinsing and blowing with compressed air failed to clean the root completely, then a clean cotton swab soaked in a solvent was applied to the root. The swab was always kept turning when it was in contact with the root to keep the polishing marks going in the desired direction. The notch roots were inspected under a low power microscope to ensure that all of the final polishing marks were more or less in the circumferential direction. After polishing, the specimens were kept in a desiccator until fatigue crack initiation tests were begun.

Apparatus

Servohydraulic testing equipment controlled by automated fatigue crack growth rate data acquisition and control software was used. A clevis and pin assembly was used at both ends of the specimen to allow in-plane rotation as the specimen was loaded. This specimen and loading arrangement was used for tension-tension loading only.

A recent survey of crack length measurement methods and their resolution [*20*] indicates that electron microscope methods are the most sensitive, followed by compliance. As electron microscope methods are not useful for continuous measurements in environments other than vacuum; the compliance method was chosen instead. A load cell and a crack mouth opening displacement (CMOD) gage were used to measure the compliance.

General Procedure

Load levels were chosen to yield data over the lifetimes of interest. Preliminary load levels were chosen on the basis of yield strength, as there is evidence that the endurance limit, at least for steels, is a function of yield strength [*7, 9*]. Runout specimens were sometimes reused by testing them at higher loads to get more data. However, they were renotched first as a crack may already have initiated undetected and because cycling at low loads before fatigue testing at high loads has been shown to increase fatigue life, a phenomenon known as "coaxing" [*21, 22*]. Renotching was designed to remove at least the plastic zone [*23*], which was estimated as $K^2/(\pi\sigma_{ys}^2)$.

The compliance of the specimens was continually monitored. An a/W increment of 0.005 was used as the definition of initiation. The initial crack length estimate was sometimes anomalous so the first stable value was used instead for computation of the target final a/W at which the software would automatically stop the test. The interval at which automated compliance data was taken was set at about one thousandth of the elapsed cycles. The intermittent nature of the crack length calculation could have allowed a crack to initiate without an indication of the corresponding cycle count. It was possible to avoid this source of uncertainty because, although the software only calculated the crack lengths intermittently, it continuously monitored the maximum crack mouth opening displacement (CMOD) and would stop the test as soon as the maximum CMOD exceeded full scale. Therefore, the CMOD gage zero offset setting was adjusted at the beginning of the test so that the output at the maximum load in the loading cycle, P_{max}, was near full scale. This enabled the software to stop the test automatically before a crack got very long even if enough loading cycles had not elapsed since the last crack length measurement to trigger another crack length calculation. A runout, or a fatigue stress level that is so low that it will effectively never initiate a crack, was defined as 10^7 cycles. The software allowed data acquisition and test stoppage based directly on cycle count so this count was entered as a final count to enable a consistent definition of a runout.

After fatigue crack initiation had been detected, the specimens were overloaded to fracture and the fracture surfaces were examined in a scanning electron microscope to yield information on the initiation sites and mechanisms.

Data was plotted with the cycles to initiation, N_i, on the abscissa and using a logarithmic scale. $\Delta K / \rho^{1/2}$ in units of stress was plotted on the ordinate using a linear scale.

Material

Aluminum alloy 7075 and titanium alloy Ti-6Al-4V were studied. A 63.5-mm thick rolled plate of over-aged 7075-T7351 was used. The chemical composition of the 7075 alloy in weight per cent supplied by the vendor is shown in Table 1. Typical tensile properties values for 7075-T7351 from a handbook [24] are tabulated in Table 2.

Table 1 - *Chemical composition (wt %) of 7075 aluminum alloy.*

Zn	Mg	Cu	Cr	Mn	Ti	Si	Fe	Al
5.70	2.52	1.59	0.20	0.05	0.04	0.09	0.17	bal

Table 2 - *Mechanical properties of 7075-T7351 aluminum alloy.*

0.2% Yield Strength, σ_{ys}, MPa	Tensile Strength, σ_{uts}, MPa	Young's Modulus, E, GPa	Elongation (51 mm G. L.) %
434	503	72	13

The Ti-6Al-4V alloy studied was in the form of a 25.4-mm thick rolled plate. Chemical analysis of the plate appears in Table 3. Procedures followed in the two heat treatments are given in Table 4. Tensile properties and plane strain fracture toughness (K_{IC}) values for each heat treatment are tabulated in Table 5. Tensile properties for the transverse (T) and longitudinal (L) orientations are provided for comparison, as an index of the degree of texture that exists for each heat-treated condition. K_{IC} values are for the TL orientation [E 399]. The MA microstructure was characterized by elongated primary alpha and the BA was acicular Widmanstatten.

Table 3 - *Chemical composition (wt %) of Ti-6Al-4V titanium alloy.*

O	Al	V	Fe	N	C	H
0.20	6.7	4.3	0.10	0.011	0.03	0.006

Table 4 - *Heat treatments of Ti-6Al-4V titanium alloy.*

Mill Anneal (MA)	(788°C/1 hr + AC), as received
Beta Anneal (BA)	(1038°C/1/2 hr + AC) + (732°C/2 hr + AC)

AC: air cool

Table 5 - *Mechanical properties of Ti-6Al-4V titanium alloy.*

Heat Treatment	MA		BA	
Fracture Toughness, K_{IC}, MPa÷m	40		76	
Orientation	T	L	T	L
0.2 % Yield Strength, σ_{ys}, MPa	1 007	948	931	898
Tensile Strength, σ_{uts}, MPa	1 034	986	1 007	982
Young's Modulus, E, GPa	130	118	130	119
Reduction in area, %	29	26	26	26
Elongation in 51 mm, %	14	15	15	15

Results and Discussion

Blunt notch wedge-opening-loaded (WOL) specimens with height $H = 63$ mm and width $W = 64.8$ mm, like the one shown in Fig. 1, were used in the fatigue crack initiation studies. The aluminum alloy specimens were 12.7 mm thick while the titanium alloy specimens were 11.4 mm thick. The blunt notches had a radius of 3.18 mm, which resulted in a stress concentration factor $K_t = 3.1$. The roots of the blunt notches were polished in the circumferential direction with the final step using 3 μm diamond paste.

Fatigue crack initiation tests were conducted at various stress intensities at a stress ratio $R = 0.10$ and a frequency of 5 Hz with a sinusoidal waveform. Although the software used was designed for fatigue crack propagation testing rather than fatigue crack initiation testing, it was easily adapted for this purpose by the proper selection of parameters. These parameters included 1) constant load amplitude fatigue loading, 2) a final normalized crack length, a/W that was 0.005 greater than the initial a/W and 3) a final cycle count of 10^7. The a/W increment was an arbitrary definition of initiation and resulted in a stress intensity factor increase of only 1% during the constant load amplitude test. Selection of a constant stress intensity factor amplitude was also possible but probably would not have had any noticeable effect on the results over this small an increment of crack growth. The final cycle count of 10^7 was an arbitrary definition of a runout.

Figures 2 and 3 are 7075-T7351 a versus N curves for, respectively, a specimen with a short lifetime and a specimen with a long lifetime. Fig. 2 is for a specimen that was polished and then pitted by 336 hours of exposure to a 3.5 % solution of sodium chloride (NaCl) in water before the fatigue test. Fig. 3 is for a specimen that was tested in the as-polished condition. The noise levels in both cases are 0.0025 mm, which is the resolution of the crack lengths indicated by the data acquisition system. Even tests run with older and noisier controllers had noise levels no higher than 0.05 mm. Initiation was arbitrarily defined as an a/W increment of 0.005, which translates to 0.324 mm for specimens of these dimensions. As this is six times as much as the highest noise level, it is clear that cracks have initiated. Scanning electron microscopy confirmed that cracks had initiated in every specimen where the data acquisition system indicated an *a/W* increment of 0.005. Although both specimens were tested at the same stress amplitude, cracks in one initiated quickly from multiple pits and the crack in the other initiated slowly from an as-polished surface. For both specimens, all of the crack growth seems to occur over the last 20 000 cycles required to meet the arbitrary definition of initiation. The curves are indistinguishable except for the number of cycles required for crack initiation. It can be seen that the amount of crack growth defined as initiation will affect the results obtained for the number of cycles to initiation. The number of cycles between the horizontal and the vertical portions of the data show that the effect of the arbitrary choice of an *a/W* increment of 0.005 as the definition of initiation on the number of cycles to initiation is less than a factor of two. Thus, it will have only a slight effect on the shape of an *S-N* curve, where the cycles to initiation are plotted on a logarithmic scale.

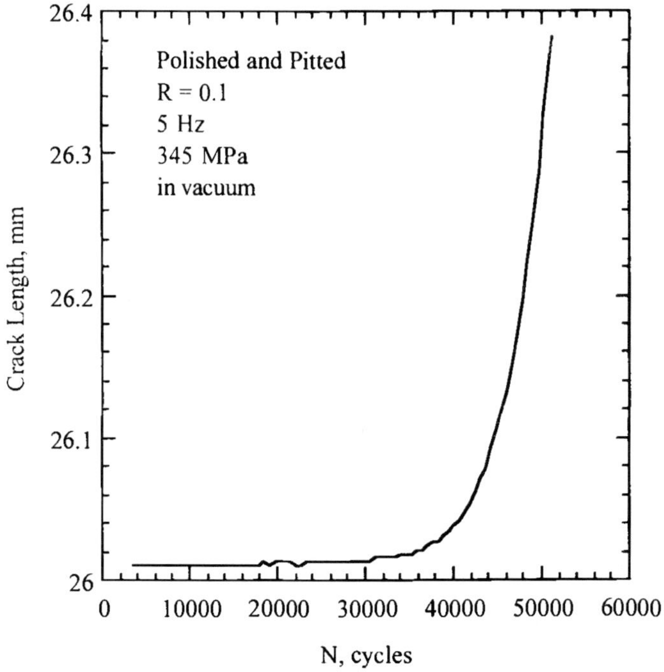

Fig. 2 - *Fatigue crack initiation curve for an aluminum alloy 7075-T7351 specimen with a short lifetime.*

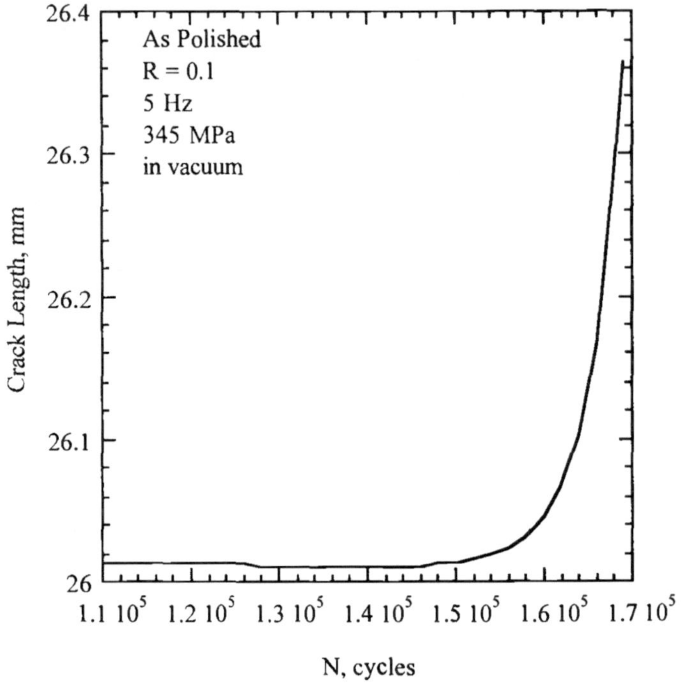

Fig. 3 – *Fatigue crack initiation curve for an aluminum alloy 7075-T7351 specimen with a long lifetime.*

Fig. 4 compares the fatigue crack initiation kinetics of S-T orientation aluminum alloy 7075-T7351 in the as-polished and in the polished-and-pitted conditions. The presence of pre-existing corrosion pits, produced by 336 hours immersion in salt water, significantly reduces the fatigue crack initiation life and threshold stress intensity of this alloy. The pre-existing pits at the blunt root surface act as stress concentration sites at which the local stresses are elevated to facilitate fatigue crack initiation. Scanning electron microscopy confirmed that the origin of the fatigue cracking in the pre-pitted specimens could be traced to these pits [*25, 26*]. This initiation mechanism demonstrates the versatility of the compliance technique for detection of crack initiation, as some other techniques are not capable of detecting changes that occur entirely within the notch root.

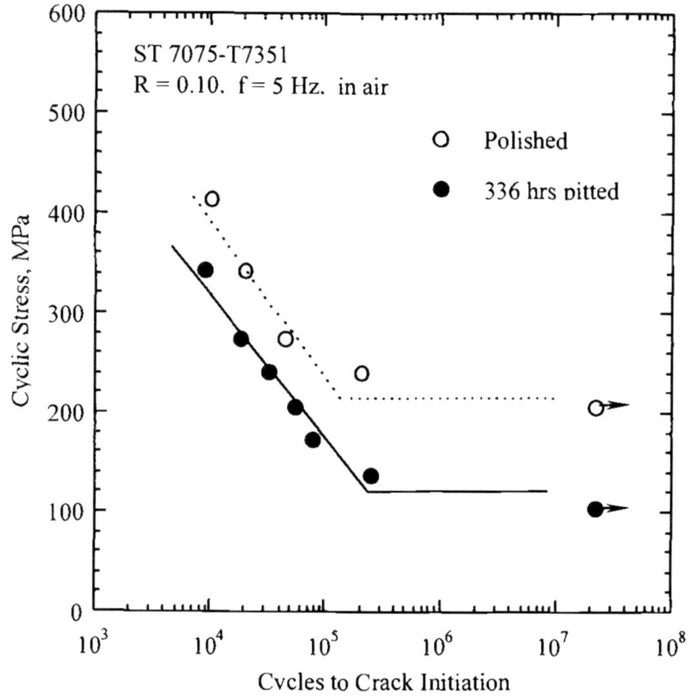

Fig. 4 – *Fatigue crack initiation kinetics of S-T orientation aluminum alloy 7075-T7351 in the as-polished and in the polished-and-pitted conditions.*

Fig. 5 shows *a* versus *N* curves for BA and MA Ti-6Al-4V specimens tested at the same stress amplitude. The MA microstructure shows significantly greater resistance to fatigue crack initiation at this stress amplitude than the BA microstructure. In addition, once a crack initiates in the MA microstructure, it reaches the *a/W* increment crack initiation criteria within a fewer number of cycles than it does in the BA microstructure. In other words, the bend shown in Fig. 5 for the MA microstructure is tighter than that for the BA microstructure. While the difference in cycles to initiation for the two microstructures could be detected by many techniques, only continuous monitoring of crack length could show differences in the sharpness of the bend in the data as the cracks begin to initiate. Previous investigations of fatigue crack growth in Ti-6Al-4V have shown that it is microstructure-sensitive, with a transition in the slope of the fatigue crack growth rate curve occurring at the point at which the reversed plastic zone size equals the average Widmanstatten packet size in beta annealed material [*27, 28*]. The exact mechanisms have been established and are being investigated.

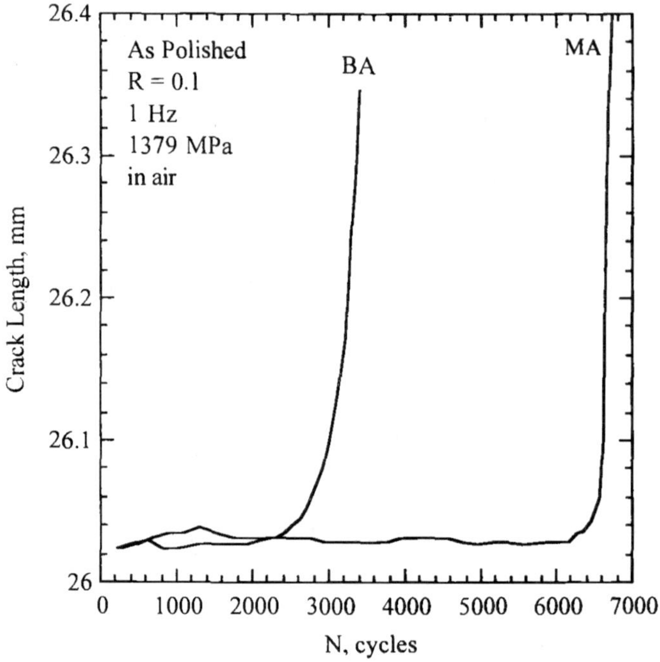

Fig. 5 – *Fatigue crack initiation curves for BA and MA Ti-6Al-4V specimens tested at the same stress amplitude.*

Summary

Commercially available software for fatigue crack growth rate testing, servohydraulic testing equipment, fracture mechanics specimens modified with blunt notches, and the use of the parameter $\Delta K / \rho^{1/2}$ were used to automate fatigue crack initiation testing. The combination of all of these elements represents a new test method. Results were presented for aluminum alloy 7075, where the effects of corrosion pits were studied and for titanium alloy Ti-6Al-4V, where the effects of heat treatment on initiation were studied. The resolution of the crack lengths indicated by the data acquisition system was 0.0025 mm. Initiation was arbitrarily defined as an *a/W* increment of 0.005, which translated to 0.324 mm of crack growth for specimens of these dimensions. The amount of crack growth defined as initiation affected the results obtained for the number of cycles to initiation. However, the uncertainty in the number of cycles to initiation was less than a factor of two and will have only a slight effect on the shape of an *S-N* curve, where the cycles to initiation are plotted on a logarithmic scale. Scanning electron microscopy confirmed that the origin of the fatigue cracking in some pre pitted specimens could be traced to pits in the blunt notch root. This initiation mechanism demonstrates the versatility of the compliance technique for detection of crack initiation, as some other techniques are not capable of detecting changes that occur entirely within the notch root. The continuous monitoring of crack length showed differences in the sharpness of the bend in the data as the cracks begin to initiate that would probably have been missed by any intermittent data acquisition technique.

References

[1] Saanouni, K. and Bathias, C., "Study of Fatigue Crack Initiation in the Vicinity of Notches," *Engineering Fracture Mechanics*, Vol. 16, No. 5, 1982, pp. 695-706.

[2] Saxena, A. and Hudak, S. J., Jr., "Review and Extension of Compliance Information for Common Crack Growth Specimens," *International Journal of Fracture*, Vol. 14, No. 5, 1978, pp. 453-468.

[3] Wessel, E. T., Clark, W. G., Jr., and Wilson, W. K., "Engineering Methods for the Design and Selection of Materials Against Fracture," DDC Report AD801001, 1966.

[4] Paris, P. C. and Shi, G. C., "Stress Analysis of Cracks," *Fracture Toughness Testing and its Applications, ASTM STP 381*, American Society for Testing and Materials, Philadelphia, PA, 1965, pp. 30-83.

[5] Wilson, W. K. and Gabrielse, S. E., "Elasticity Analysis of Blunt Notched Compact Tension Specimens," unpublished Westinghouse Research Laboratories data.

[6] Jack, A. R. and Price, A. T., "The Initiation of Fatigue Cracks From Notches in Mild Steel Plates," *International Journal of Fracture*, Vol. 6, No. 4, Dec. 1970, pp. 401-409.

[7] Barsom, J. M. and McNicol, R. C., "Effect of Stress Concentration on Fatigue-Crack Initiation in HY-130 Steel," *Fracture Toughness and Slow Stable Cracking, ASTM STP 559*, P. C. Paris and G. R. Irwin, Eds., American Society for Testing and Materials, West Conshohocken, PA, 1974, pp. 183-204.

[8] Clark, M. G., Jr., "Evaluation of the Fatigue Crack Initiation Properties of Type 403 Stainless Steel in Air and Steam Environments," *Fracture Toughness and Slow Stable Cracking, ASTM STP 559*, P. C. Paris and G. R. Irwin, Eds., American Society for Testing and Materials, West Conshohocken, PA, 1974, pp. 205-224.

[9] Rolfe, S. T. and Barsom, J. M., *Fracture and Fatigue Control in Structures*, Prentice-Hall, Englewood Cliffs, NJ, 1977, pp. 208-231.

[10] Dowling, N. E., Brose, W. R., and Wilson, W. K., "Notched Member Fatigue Life Predictors by the Local Strain Approach," *Fatigue Under Complex Loading - Analysis and Experiments*, Society of Automotive Engineers, Warrendale, PA, 1977, pp. 55-84.

[11] May, R. A., Stuber, A., and Rolfe, S. T., "Effective Utilization of High Yield Strength Steels in Fatigue," *Welding Research Council Bulletin 243*, New York, November 1978, pp. 1-26.

[12] Braglia, B. L., Hertzberg, R. W., and Roberts, R., "Crack Initiation in a High-Strength Low-Alloy Steel," *Fracture Mechanics, ASTM STP 677*, C. W. Smith, Ed., American Society for Testing and Materials, Philadelphia, PA, 1979, pp. 290-302.

[13] Dowling, N. E., "Fatigue at Notches and the Local Strain and Fracture Mechanics Approaches," *Fracture Mechanics, ASTM STP 677*, C. W. Smith, Ed., American Society for Testing and Materials, Philadelphia, PA, 1979, pp. 247-273.

[14] Yoder, G. R., Cooley, L. A., and Crooker, T. W., "A Comparison of Microstructural Effects on Fatigue Crack Initiation and Propagation in Ti-6Al-4V," AIAA 82-0660-CP, *Proceedings of AIAA/ASMEASCE/AHS 23rd Structures, Structural Dynamics and Materials Conference*, American Institute of Aeronautics and Astronautics, New York, NY, 1982, pp. 132-136.

[15] Hertzberg, R. W., *Deformation and Fracture Mechanics in Engineering Materials*, 4th Ed., Wiley, New York, 1995, pp. 573-575.

[16] Novak, S. R., "Corrosion Fatigue Crack Initiation Behavior of Four Structural Steels," Corrosion Fatigue: Mechanics, Metallurgy, Electrochemistry, and Engineering, ASTM STP 801, T. W. Crooker and B. N Leis, Eds., American Society for Testing and Materials, Philadelphia, PA, 1983, pp. 26-63.

[17] Prater, T. A. and Coffin, L. F., "The Use of Notched Compact-Type Specimens for Crack Initiation Design Rules in High-Temperature Water Environments," *Corrosion Fatigue: Mechanics, Metallurgy, Electrochemistry, and Engineering, ASTM STP 801*, T. W. Crooker and B. N Leis, Eds., American Society for Testing and Materials, Philadelphia, PA, 1983, pp. 423-444.

[18] Yoder, G. R., Cooley, L. A., and Crooker, T. W., "Observations on a Fracture Mechanics Approach to Fatigue Crack Initiation in Ti-6Al-4V," *Fracture Mechanics, ASTM STP 868*, M. F. Kanninen and A. T. Hopper, Ed., American Society for Testing and Materials, Philadelphia, PA, 1985, pp. 392-405.

[19] Provan, J. W., Bonsang, F., Rezai, K., and Powell, A., "A Combined Fatigue Crack Initiation and Propagation Specimen – its Initial Development," *Theoretical and Applied Fracture Mechanics*, Vol. 15, 1991, pp. 115-129.

[20] *ASM Handbook*, Volume 19, ASM International, Materials Park, OH, 1996, p. 210.

[21] Forrest, P. G., *Fatigue of Metals*, Pergamon, Addison - Wesley, 1962, pp. 119-120 and 344-345.

[22] Knott, J. F., *Fundamentals of Fracture Mechanics*, Butterworths, 1973, pp. 235-236.

[23] Broek, D., *Elementary Engineering Fracture Mechanics*, Sijthoff and Noordhoff, 1982, pp. 92-93

[24] Anon., *Aluminum Standards and Data*, Aluminum Association, 2000, pp. 2-3 and 2-9.

[25] Pao, P. S., Gill, S. J., and Feng, C. R., "On Fatigue Crack Initiation From Corrosion Pits in 7075-T7351 Aluminum Alloy," *Scripta Materiala*, Vol. 43, No. 5, 2000, pp. 391-396.

[26] Pao, P. S., Feng, C. R., and Gill, S. J., "Corrosion-Fatigue Crack Initiation in 7075 and 7050 Aluminum Alloys," *Corrosion*, Vol. 56, No. 10, pp. 1022-1031, Oct 2000.

[27] Irving, P. E., and Beevers, C. J., "Microstructural Influences on Fatigue Crack Growth in Ti-6Al-4V," *Materials Science and Engineering*, Vol. 14, 1974, pp. 229-238.

[28] Yoder, G. R., Cooley, L. A., and Crooker, T. W., "Observations on Microstructurally Sensitive Fatigue Crack Growth in a Widmanstatten Ti-6Al-4V Alloy," *Metallurgical Transactions A*, Vol. 8A, 1977, pp. 1 737-1 743.